"十四五"职业教育部委级规划教材

教育部国家职业教育专业教学资源民族文化传承与创新子库"中国
丝绸技艺民族文化传承与创新"配套双语教材

江苏省高等职业院校高水平专业群"纺织品检验与贸易"配套教材

中华名锦
Chinese Famous Brocades

蒋秀翔　周　燕 ◎主　编

吴惠英　原海波　徐超武 ◎副主编

马娟娟 ◎译

U0162877

中国纺织出版社有限公司

内容提要 /Summary

蜀锦、宋锦、云锦是中国的三大名锦，本书介绍了三大名锦的起源和发展、品种和花型，归纳了中国古代、近代和现代的织锦技术，总结了蜀锦、宋锦和云锦的传承与创新，传扬了中国自古以来璀璨的丝绸文化。本书使读者能充分领略中国历代工匠的精湛技艺，感受他们精益求精的工匠精神，折服于中国灿烂辉煌的丝绸文化。本书可供研究生、本科及专科学生研究使用。

Shu brocade, Song brocade and Yun brocade are the three famous brocades in China. This book introduces the origin, development, varieties and patterns of the three famous brocades, summarizes the ancient and modern brocade weaving techniques in China, generalizes their inheritance and innovation, and displays the splendid silk culture of China since ancient times. Readers can fully appreciate the exquisite skills of Chinese craftsmen of all ages, feel their craftsmanship of striving for perfection, and be impressed by Chinese brilliant silk culture. The book is suitable for the research of postgraduate students, undergraduate students and junior college students.

图书在版编目（CIP）数据

中华名锦 =Chinese Famous Brocades：汉英对照 / 蒋秀翔，周燕主编；吴惠英，原海波，徐超武副主编；马娟娟译 . -- 北京：中国纺织出版社有限公司，2023.4

"十四五"职业教育部委级规划教材　教育部国家职业教育专业教学资源民族文化传承与创新子库"中国丝绸技艺民族文化传承与创新"配套双语教材　江苏省高等职业院校高水平专业群"纺织品检验与贸易"配套教材

ISBN 978-7-5229-0077-3

Ⅰ . ①中… Ⅱ . ①蒋… ②周… ③吴… ④原… ⑤徐… ⑥马… Ⅲ . ①锦—介绍—中国—高等职业教育—教材—汉、英 Ⅳ . ①TS146

中国版本图书馆 CIP 数据核字（2022）第 217999 号

ZHONGHUA MINGJIN

责任编辑：孔会云 沈 靖　责任校对：高 涵　责任印制：王艳丽

中国纺织出版社有限公司出版发行
地址：北京市朝阳区百子湾东里 A407 号楼　邮政编码：100124
销售电话：010—67004422　传真：010—87155801
http://www.c-textilep.com
中国纺织出版社天猫旗舰店
官方微博 http://weibo.com/2119887771
北京通天印刷有限责任公司印刷　各地新华书店经销
2023 年 4 月第 1 版第 1 次印刷
开本：787×1092　1/16　印张：11.5
字数：250 千字　定价：88.00 元

前　言 / Foreword

　　丝绸是中国古老文化的象征，对促进人类文明的发展做出了不可磨灭的贡献。随着国家"一带一路"倡议的推进，中国丝绸以其卓越的品质、精美的花色和丰富的文化内涵闻名于世。锦起源于中国，在我国发展的历史已有3000多年。

　　Silk is a gem of ancient Chinese culture and has made an indelible contribution to the development of human civilization. With the promotion of "the Belt and Road Initiative", Chinese silk is world-renowned for its excellent quality, exquisite colors and rich cultural connotations. Brocade originated in China and can be traced back to more than 3000 years ago.

　　发掘传承千年的丝绸文化，弘扬中华璀璨文明，在国家"一带一路"倡议下显得尤为重要和迫切。本书分别阐述了中国三大名锦——蜀锦、宋锦和云锦的起源、发展、品种、花色以及传承，介绍了古代、近代和现代三个时期锦的织造技艺，供广大纺织专业技术人员和丝绸文化爱好者学习和参考。

　　It is particularly important and urgent to explore and inherit the silk culture for thousands of years and promote the splendid civilization of China today when "the Belt and Road Initiative" is advocated. This book expounds the origin, development, varieties, colors and inheritance of the three famous brocades in China, including Shu brocade, Song brocade and Yun brocade according to the order of their appearance in history, and introduces the brocade weaving craftsmanship in ancient and modern times. The book can be used for reference by professional technicians and silk culture enthusiasts.

　　全书共四章。第一章由成都纺织高等专科学校原海波编写；第二章由苏州经贸职业技术学院吴惠英编写；第三章由苏州经贸职业技术学院周燕编写；第四章第一节由苏州经贸职业技术学院徐超武编写，第二节和第三节由苏州经贸职业技术学院蒋秀翔编写。全书由蒋秀翔统稿，全书英文由西安工程大学的马娟娟翻译，胡伟华总译审。

　　This book consists of four chapters. Chapter 1 is compiled by Yuan Haibo from Chengdu Textile College; chapter 2 is compiled by Wu Huiying from Suzhou Economic and Trade Vocational and Technical College; chapter 3 is compiled by Zhou Yan from Suzhou Economic and Trade Vocational and Technical College. As for chapter 4, section I is compiled by Xu Chaowu of Suzhou Economic and Trade Vocational and Technical College, and section Ⅱ and Ⅲ are compiled by Jiang Xiuxiang from Suzhou Economic and Trade Vocational and Technical College. The final compiler and editor is Jiang Xiuxiang, the translator is Ma Juanjuan of Xi'an Polytechnic University, and the chief professor of translation is Hu Weihua.

因笔者水平有限，疏漏之处在所难免，恳请广大读者批评指正。

Due to the limited knowledge of the author, the shortcomings and mistakes of this book are inevitable. Criticism and suggestions are welcome.

编者

2021年10月

翻译前言
Translation Preface

《中华名锦》是教育部国家职业教育专业教学资源民族文化传承与创新子库"中国丝绸技艺民族文化传承与创新"配套双语教材中的一种，也可供对纺织服装感兴趣的人士及纺织服装行业从业者学习参考。本书英译部分旨在引发国内外纺织服装领域从业者的兴趣，为中国纺织服装业的国际拓展发挥作用。

Chinese Famous Brocades is one of the Bilingual teaching materials "Inheritance and Innovation of Chinese Silk Skills and National Culture", a sub-library of national vocational education professional teaching resources of the Ministry of Education. It can also serve as a reference for the readers who are interested in textiles and clothing as well as practitioners in the textile and clothing industry. The English version of the course will be of interest to many practitioners in the industry at home and abroad. It will play an important role in the international expansion of the Chinese textile and clothing industry.

本书由西安工程大学马娟娟及其团队翻译完成。全书共四章，第一、第二、第四章由马娟娟翻译；第三章由西安工程大学翻译硕士刘晗裕翻译。刘晗裕协助校对了全书英文。

It was translated by Ma Juanjuan and her team at Xi'an Polytechnic University. The book consists of four chapters, Ma Juanjuan undertook the translation of the first chapter, the second chapter, the fourth chapter; Liu Hanyu, who is a MTI (Master of Translation and Interpreting) student of Xi'an Polytechnic University, undertook the translation of the third chapter. Liu Hanyu also helped with the proofreading of the English version of the four chapters.

西安工程大学的外籍教师道格拉斯·劳伦斯对全书中的英文进行了审阅，并提出了许多宝贵的修改建议。在此衷心感谢他的辛勤工作！

Douglas Laurence, a foreign teacher of Xi'an Polytechnic University, reviewed all the translations and put forward valuable suggestions for revision. We are grateful for his hard work!

译者
2022年11月

目 录/Contents

○ 第一章

蜀锦
Shu Brocade

中国是丝绸的发源地，在大约 7000 年前（新石器中早期）就发明了桑蚕丝织技术，从西周时（前 1046 ~ 前 771 年）就有织锦。巴蜀是中国丝绸文化发源地之一，且巴蜀特殊的地理位置，自古就盛产桑蚕，素有"蚕丛古国"之称。

China is the birthplace of silk. Mulberry silk weaving technique was invented about 7,000 years ago（early Neolithic Age）, and brocade came into being since the Western Zhou Dynasty （1046BC ~ 771BC）. Bashu（two ancient cities located in today's Sichuan）is one of the birthplaces of Chinese silk culture, and with its special geographical location, has been rich in silkworms since ancient times, and has been known as the "ancient silk country".

蜀锦是蜀地的提花丝织物，因产于蜀地而得名，传统丝织工艺生产锦缎的历史悠久，影响深远，是巴蜀丝绸文化的代表。从春秋战国伊始，兴于汉、盛于唐，至今已有近 3000 年的历史。

Shu brocade is a jacquard silk fabric, which is named after being produced in Shu（roughly in today's Sichuan Province）. It has a long history and far-reaching influence on the production of traditional silk brocade, and represents the Bashu silk culture. It emerged during the Spring and Autumn Period and Warring States Period, flourished in the Han and Tang Dynasty. Up until now, it has a history as long as nearly 3,000 years.

蜀锦是巴蜀丝绸文化的代表，对丝绸织锦业的发展有着广泛而深刻的影响，并在我国传统工艺美术史上中留下了辉煌的一页。蜀锦以其精湛的技艺、生动的图案、艳丽的色彩、独特的组织、精致细腻的质地，被誉为"东方瑰宝""中华一绝"。本章主要阐述了蜀锦在各个不同历史阶段的演变和发展，详述了蜀锦的类别、花色以及传承和发展。

Shu brocade represents the silk culture in Bashu（two ancient cities located in today's Sichuan）, which has a wide and profound impact on the development of silk brocade industry and left a brilliant page in the history of Chinese traditional arts and crafts. Shu Brocade, with its exquisite skills, vivid patterns, gorgeous colors, unique fabric weaves and delicate textures, is known as "the oriental gem" and "a marvel in China". This chapter mainly expounds the evolution and development of Shu brocade in different historical stages, and details its categories, colors as well as its inheritance and development.

第一节　蜀锦的起源与发展/The Origin and Development of Shu Brocade

蜀地常年气候湿润、土壤肥沃。成都平原的地势条件更是农耕养蚕的绝佳条件，这为古蜀先民发现野蚕后进行家养的驯化提供了非常好的条件。古蜀先民通过长期养蚕驯蚕，发展抽丝剥茧技艺，为蜀锦的出现奠定了基础。

Shu has humid climate all the year round and fertile soil. Moreover, the topography of Chengdu Plain is an excellent condition for farming and sericulture. These are all beneficial for domestication of wild silkworms after the ancestors of ancient Shu discovered them. The ancient Shu ancestors laid the foundation for the emergence of Shu brocade through the long-term development of sericulture, silkworm training and cocoon stripping skills.

在新石器时期的渔猎时代，四川西部岷江上游，居住着以从事驯养蚕及将茧抽丝为主要职业的氏族，称为"蚕丛"。蚕丛氏南迁成都后，教民养蚕，为巴蜀养蚕治丝、织绢锦孕育了条件。

During the fishing and hunting era of Neolithic Age, in the upper stream of Minjiang River in western Sichuan, there lived a clan called "Cancong" who was engaged in domestication of silkworms and reeling off raw silk from cocoons. After Cancong Clan moved south to Chengdu, they started to teach people to raise silkworms, which provided conditions for sericulture and silk weaving in Bashu area.

根据东晋史学家常璩在《华阳国志·巴志》中关于蜀地织物的描述："禹会诸侯于涂山，执玉帛者万国，巴蜀往焉。"可见距今4000多年前的巴蜀已能生产丝织品"帛"了。再从三星堆出土的商代石器、陶器、青铜器等不同材质的纺轮，可推断蜀地在商代就能纺制不同规格的丝线，进行绣花与丝织。

Chang Qu, a historian of the Eastern Jin Dynasty, once described the fabric of Shu in *The Records of Ba* of *Records of Huayang Kingdom* as follows: "Yu the Great gathered all the governors of vassal states in Tushan, and governors from numerous vassal states, including Ba and Shu, were present at the court, with jade and Bo as tributes." According to this, Bashu was able to produce Bo, a kind of white silk woven strip mainly used for hand writing and painting, about more than 4,000 years ago. From the spinning wheels discovered in Sanxingdui Ruins of different materials, such as stone, pottery and bronzes, it can be inferred that silk threads of different specifications can be spun for embroidery and silk weaving in Shu during Shang Dynasty.

春秋战国时期，蜀地的蚕桑丝织业持续发展，提花织锦技艺也日臻完善，已向朝廷赋税纳贡。由于蜀地丝绸质优、量大、品种多，在公元前4世纪，甚至更早的商代中晚期就已初步形成了"南方丝绸之路"，由"蜀身毒道"把蜀地的丝织品和其他货物销往印度、缅甸又转运中亚。

During the Spring and Autumn Period and the Warring States Period, the sericulture and silk weaving industry in Shu continued to develop, and the jacquard brocade skills were gradually improving. People had started to paying taxes and tributes to the imperial court. Because silk in Shu boasted excellent quality, high production and rich varieties, as early as in the 4th century BC, or even earlier in the middle and late Shang Dynasty, Shu–Hindu Road, also known as "Southern Silk Road", was initially formed. The silk fabrics and other goods produced in Shu were sold to India and Myanmar and then transferred to Central Asia through this route.

战国后期，秦始皇统一六国之际，巴蜀以"丝绵锦帛"资助了秦军军费，周慎靓王定六年秋，秦惠文王征服巴蜀，修建成都城，在城南夷里桥南岸筑"锦官城"，把织锦工人集中生产，设"锦官"履行税收和生产监管，可见，当时蜀地丝织业已成为一项重要产业。

When Emperor Qin Shihuang unified the six states in the late Warring States Period, Bashu funded Qin army with silk and brocade as military expenditure. In 315 BC, after King Huiwen of Qin conquered Bashu, Chengdu City was constructed. "Jinguan City" on the south bank of Yili Bridge in the south of the city was built to concentrate the production of brocade workers. An official post named "Jinguan", which literally means "brocade officer", was set up to perform the duty of taxation and production supervision. So we can infer that silk weaving in Shu had become an important industry at that time.

西汉郫县即蜀郡成都人扬雄在《蜀都赋》中说："若挥锦布绣，望芒兮无幅。尔乃其人，自造奇锦。"可见西汉成都织锦生产之盛和蜀锦用途之广。

Yang Xiong, a local poet of Shu once described the splendid silk weaving scene here in his poem *Odes to Shu*, which shows that the brocade production in Chengdu during the Western Han Dynasty was flourishing and Shu brocade had been widely used then.

汉武帝时，成都生产的织锦品种花色繁多，丝织业繁盛。到东汉时，成都已成为与齐鲁临淄、陈留、襄邑并驾齐驱的织锦中心。

During the reign of Emperor Wu of the Han Dynasty, producing wide varieties of brocades and promoted the flourishing of silk weaving industry. By the Eastern Han Dynasty, Chengdu, together with Linzi, Chenliu and Xiangyi, became the brocade weaving centers.

三国时，陈留、襄邑属于魏国，所产织锦有"魏锦"之称，成都织锦则称为"蜀锦"。刘备占领益州，打开刘璋仓库，发现丝织品库存很大。东汉末年，诸葛亮为完成统一大业，极力提倡蚕织："今民贫国虚，决敌之资，唯仰锦耳。"由此蜀汉丝绸生产又有很大发展。

During the Three Kingdoms Period, the brocade produced in Chenliu and Xiangyi under the jurisdiction of Wei State was called "Wei brocade", while the brocade produced in Chengdu "Shu brocade". After Liu Bei occupied Yizhou, when he opened the warehouse, he found that there was a large stock of silk fabrics. At the end of the Eastern Han Dynasty, Zhuge Liang strongly advocated silkworm weaving in order to accomplish the reunification, "Since the people are poor and the nation's coffer is empty, we should advocate brocade weaving to raise the military expenditure". Thus the silk production in Kingdom of Han kept on flourishing.

诸葛亮在成都设置的锦官，是专业管理蜀锦生产的官吏，在成都的东南角，自古就是盛产蜀锦的地方，政府在此设官管理，故称"锦官城"。在锦官城下，有一条"流江"的水流过，这条流江后来被人们称为"锦江"或"濯锦江"。锦官城上的楼称为"锦江楼"，锦工们居住的地方被称为"锦里"和"濯锦厢"，成都则被称为"锦市"。蜀锦的名声已呈压倒之势，远远超过陈留、襄邑生产的魏锦。

"Jinguan", the brocade officer, nominated by Zhuge Liang, specialized in managing the production of Shu brocade. "Jinguan City", located in the southeast corner of Chengdu, had been rich in Shu brocade since ancient times. At the gate of Jinguan, there was a river called "Liujiang River", which was also called "Jinjiang River" or "Zhuojin River" later. The place where brocade workers lived was called "Jinli" (brocade street) and "Zhuojinxiang" (brocade rinsing district), and Chengdu was called "Jinshi" (brocade market). The fame of Shu brocade had overwhelmingly exceeded that of Wei brocade produced by Chenliu and Xiangyi during the Three Kingdoms Period.

魏晋南北朝时期，蜀锦的发展并未因为国家的动荡而停滞不前，相反因蜀地偏，远蜀锦的发展并未受到影响，且比中原地区的织锦发展更为繁盛。曾有刘宋时期的一位郡守山谦之在文献《丹阳记》中记载："江东历代尚未有锦，而成都独称妙。"

During the Wei, Jin, Southern and Northern Dynasties, the development of Shu brocade was not prevented by chaos and wars. On the contrary, because of the remoteness of Shu, it was more prosperous than that in the central plains. Shan Qianzhi, a county magistrate in Liu Song Dynasty, once recorded in the document *A Chorography of Danyang*: "There was no brocade on the south of Changjiang River in the past dynasties, but the brocade produced in Chengdu can be called a wonder alone."

隋时，成都"水陆所凑，货殖所萃"，织造的绫锦，质量精美"侔于上国"。

In the Sui Dynasty, Chengdu was the hub of water and land transportation and so became the trade center, and the brocade produced here took the lead in the whole country with its exquisite quality.

唐代在织造技艺和图案设计上的进步对整个丝织手工业有着重要的影响，这一进步改革了织机，让蜀锦步入纬锦时代，还在纹样上融入了外来的文化元素，创造出了闻名世界的陵阳公锦，不仅使唐朝蜀锦织锦的面料、纹样更加精美，更是让蜀锦和大唐盛世一起闻名海外。目前海外很多国家的博物馆都有收藏来自大唐盛世的蜀锦。著名的唐代文学家刘禹锡在一首诗中对蜀锦的描述堪称赞美蜀锦之经典："濯锦江边两岸花，春风吹浪正淘沙。女郎剪下鸳鸯锦，将向中流匹晚霞。"

During the Tang dynasty, the much-improved weaving skills and pattern design had an important influence on the whole silk handicraft industry. The loom was reformed and Shu brocade entered the weft brocade era. The world-famous Linyanggong Pattern was created by incorporating some foreign cultural elements. It not only made the fabrics and patterns of Shu brocade of the Tang Dynasty more exquisite, but also made it well-known overseas. At present, museums in many countries have collections of Shu brocade of the Tang Dynasty. Liu Yuxi, a famous poet in the Tang Dynasty, described Shu brocade's beauty in one of his poems as follows: "Flowers blossom along both sides of the ZhuoJinjiang River, and the spring breeze blows the waves to wash the sand. When the girl cuts off the mandarin-duck-pattern brocade, it glows in the midstream with the brilliant sunset."

五代十国时期，虽然社会处于大分裂状态，但是位于蜀地的蜀主依然在大力推动蜀锦的发展。元代费著在《蜀锦谱》中记载，宋元时期蜀锦的织造机器和技艺越来越成熟，纹样的设计上既继承于唐代蜀锦，又因为审美风格的变化，在面料和纹样的设计风格上新增了不同于从前的素雅风格，例如有名的六答晕蜀锦、落花流水锦等，这让宋代时期的蜀锦既有唐代时期繁花似锦的风格，又有淡雅如兰的情调。

During the period of Five Dynasties and Ten Kingdoms, although the contemporary society was in a divided state, the rulers of Shu were still vigorously promoting the development of Shu brocade. During the Yuan Dynasty, a book titled *Shu Brocade Manual*, written by Fei Zhu, recorded that both the weaving machines and weaving skills of Shu Brocade during the Song and Yuan Dynasties had been quite mature. Its pattern designs, inherited from that of the Tang Dynasty, due to the aesthetic change, had been greatly enriched by adding some more fabrics and pattern designs of simple but elegant styles, such as the famous Liudayun pattern design, dropping petals and flowing water pattern design, which made Shu brocade in the Song Dynasty not only had the complicated and flowery beauty in the Tang Dynasty, but also had the quietly elegant appeal.

宋元时期的蜀锦是先盛后衰，而明清时期的蜀锦则是先衰后盛。明代初期蜀锦的发展受到元代的影响，全国织造中心已经集中在江南一带，各类文献和出土的蜀锦织物都可以看出当时的蜀锦织锦作坊数量和规模都完全被限制，蜀锦的织造技艺和生产也大大受到了打击。

Shu brocade in the Song and Yuan Dynasties flourished first and then declined, while that in the Ming and Qing Dynasties declined first and then flourished. The development of Shu brocade in the early Ming was influenced by the Yuan Dynasty, and the national weaving center was concentrated in the south of the Yangtze River. From historical documents and unearthed Shu brocade fabrics, it could be inferred that the number and scale of Shu brocade workshops at that time were restricted, and the weaving skills and production were severely hampered.

明末清初，蜀地战乱长达37年。蜀地的丝织业已经摧残殆尽。自康熙起，清初外逃或被掳锦工，又回到成都，重操旧业，锦城又响起了"轧轧弄机声"。雍正年间，外地锦工入迁来到蜀地，在重庆、璧山等地张机织锦，促进了丝织业的恢复，但因基础破坏大，恢复起来十分缓慢。

In the late Ming and early Qing Dynasties, the war in Shu lasted for 37 years. The silk weaving industry in Shu was destroyed. During the reign of Qing Emperor Kangxi, the brocade workers who had fled or been captured to other places in the early Qing Dynasty returned to Chengdu and resumed their former career. The sound of weaving machines could be heard again. During the reign of Qing Emperor Yongzheng, brocade workers, who moved to Shu from other places, started to weave brocade in Chongqing, Bishan and other places, which helped to recover silk weaving industry. However, due to great damage before, the recovery was very slow.

到明末清初, 蜀地遭受了几十年的战乱, 整个蜀地处于民不聊生的状态, 直到清代中期, 蜀地社会状态逐渐稳定, 官方大力扶持蜀地织造业的发展, 工人们逐渐回到蜀地, 蜀地的织造业才慢慢回春。后蜀锦的发展随着社会的审美和先进织造技艺的革新, 还出现了著名的晚清三绝锦。

At the end of Ming Dynasty and the beginning of Qing Dynasty, Shu suffered decades of war, and people in the whole Shu lived in misery. Until the middle of Qing Dynasty, the social state of Shu gradually stabilized. Because the government vigorously supported the weaving industry, the workers gradually returned to Shu, and the local weaving industry slowly recovered. During the late of Qin Dynasty, with the innovation of social aesthetics and weaving skills, three wonders of Shu brocade with lasting charm in the late of Qing Dynasty were produced, including Yuehua (moonlight), Yusi (drizzle) and Fangfang (patterns in squares).

辛亥革命前后, 禁止民间穿绸着缎和不准用玄黄色的 "衣禁" 被取消, 团花马褂和锦缎鞋帽风行一时, 四川出现了丝织业的 "黄金时代"。成都的织锦业也随着织造及染色技艺的提高, 在传统技艺的基础上, 有了新式的纹样设计, 技艺达到炉火纯青的地步, 生产出了 "月华" 锦、"雨丝" 锦、"方方" 锦流芳百世的 "晚清三绝"。

Before and after the 1911 Revolution, the "clothing ban", which prohibited people from wearing silk fabrics and dark yellow, was abolished. Round-pattern mandarin jacket, brocade shoes and hats were popular for a while, and the golden age of silk weaving industry appeared in Sichuan. With the improvement of weaving and dyeing skills, new pattern designs were created on the basis of traditional skills, and the brocade weaving skills have reached perfection, "Yuehua brocade" "Yusi brocade" and "Fangfang brocade", three unique brocade skills in late Qing Dynasty were created at the same time.

民国初年, 成都开业机房仅350余家, 织机971台, 从业1712人, 年产绸缎绫锦48000多尺; 抗日战争爆发后, 沿海地区绸厂内迁, 促进成都丝织业有一定发展。抗日战争胜利后, 国民党发动内战, 物价一涨再涨, 美国货充斥市场, 川绸无力竞争, 成都机房90%停产, 蜀地丝织业濒临毁灭边缘。

In the early years of the Republic of China, there were only 350 workshops, 971 looms and 1,712 employers in Chengdu, with an annual output of only about 48,000 chi (a unit of length, about 1/3 meter） of silks and brocade. While after the outbreak of the Anti-Japanese War, silk factories in coastal areas moved to inland cities, which promoted silk weaving industry in Chengdu to a certain extent. After the victory of the Anti-Japanese War, the Kuomintang Party launched a civil war, and prices jumped again and again. American cloth flooded the market, and Sichuan silk was unable to compete, 90% of silk workshops in Chengdu had shut down, and the silk industry in Sichuan was on the brink of destruction.

从1951年起，为帮助丝织业发展生产，政府鼓励各地丝织户组织起来，产品由政府统购包销。当年5月13日，成都部分失业工人组建"成都市丝织业工人临时自救工厂"，有木机18台，在青羊宫瘟祖庙开业。9月28日，改组为"成都市丝织生产合作社"，使其恢复与发展蜀锦生产。

Since 1951, in order to recover the silk industry, silk weavers from all over the country were encouraged to organize, and the products were purchased and underwritten by the government. On May 13th of that year, "Temporary Self-help Factory for Silk Weaving Workers in Chengdu", set up by some unemployed silk workers in Chengdu, opened in Wenzu Temple of Qingyang Taoist Temple, only with 18 wooden looms. On September 28th, it was reorganized into "Chengdu Silk Production Cooperative" to resume and develop Shu brocade production.

1956年，经改组调整为15个生产合作社，职工共1753人。成都市丝织合作社于1958年8月改社建厂，并于当年底一度转为地方国营。1962年6月29日，除民康染厂、成都丝绸厂外，全部退回手工业合作工厂。20世纪70年代以前，丝织企业很少改造，发展缓慢。

In 1956, it was reorganized and adjusted into 15 production cooperatives with 1,753 employees. The Silk Weaving Cooperative in Chengdu changed into a factory in August 1958, and once turned into a local state-owned company at the end of that year. On June 29th, 1962, except Minkang Dyeing Factory and Chengdu Silk Factory, all of them were returned to handicraft cooperative factories. Before 1970s, silk weaving enterprises were rarely reformed and developed slowly.

1980年以后，随着改革开放，国家投资与自筹资金相结合，进行技术改造，丝织业生产得到较快的发展。截至1988年，成都拥有12家丝织企业和2085台织机，100多个品种，年产量1000万米，从业人员达9300多人。

After 1980, with the reform and opening up, the combination of state investment and self-raised funds, as well as the technical transformation, the silk weaving industry developed rapidly. By 1988, there had been 12 silk weaving enterprises in Chengdu, with 2,085 looms and more than 9,300 employees. Their annual output could amount to 10 million meters, of more than 100 varieties.

20世纪90年代后期，人们对织锦类产品的需求不再是以保暖性的服装和被面类为主，而是能体现历史文化底蕴的丝绸艺术品、工艺装饰品和收藏品等。因此，蜀锦大部分作为装饰品，用作房间壁画、沙发摆件等，深受人们的喜爱。

Since the late 1990s, what the brocade consumers demand were the silk artworks for decoration and collection. Shu brocade were mainly used as decorations, such as wall decorations, sofa ornaments, etc., highly popular among users.

实用品也是现代蜀锦中运用最为广泛的，渗入各行各业，一个零钱袋、一个名片盒、一个笔记本，都可借助蜀锦提升其实用价值。

Shu brocade are also been used for articles for daily use, such as wallets, card cases and even notebooks, whose practical value are all enhanced through the embellishment of Shu brocade.

作为收藏品，蜀锦所包含的艺术价值更高，此类产品多体现在传统手工蜀锦领域。传统蜀锦工艺繁复、耗时耗工，可谓"寸锦寸金"。如唐代传入日本的"四天纹锦"，采用传统工艺复制的该纹样织锦，因其具有丰富的历史文化内涵及采用现代罕有的手工工艺，故而拥有很高的收藏价值。

Shu brocade, especially those traditional hand-made ones, as collections, have higher artistic value. Traditional Shu brocade, made with much more complicated technology, are much more time-consuming and labor-consuming, which is described as "an inch of brocade values an inch of gold". For example, brocade with patterns of four hunting heavenly kings, which were introduced into Japan in the Tang Dynasty, is reproduced by traditional techniques today. The reproduced brocade also has high collection value because of its rich historical, cultural connotation and rare manual handicrafts.

随着科技的进步，蜀锦织造技艺也获得了持续的发展。纹织CAD辅助系统可代替并简化意匠的绘制、编制组织，存储投梭、换道、纹板等信息，极大地提高了纹制工作效率。同时，又将数码技术引入传统手工挑花和织锦技艺，实现了电子挑花、电子提花，使无纹板设计提花织造一体化，使产品设计更为快捷。因此，传统蜀锦织造的生产作用已渐渐被文化保护所代替，也表示蜀锦进入新的发展纪元。纵观蜀锦的发展兴衰，其经历了历史的繁盛与沧桑，贯穿了整个蜀地文明的发展。

With the progress of science and technology, the weaving technology of Shu brocade has also achieved continuous development. The Jacquard Weaving CAD System can replace and simplify the drawing and weaving of designers, store the information of picking, lane changing and pattern board, which greatly improves the efficiency of jacquard weaving. At the same time, the digital technology realizes the cross-stitch and brocade weaving by machine, achieves electronic cross-stitch and jacquard weaving without pattern board, and makes product design efficient and quick. Therefore, the producing function of traditional Shu brocade has been gradually replaced by cultural protection, which also indicates that Shu brocade has entered a new era of development. Looking at the rise and fall of Shu brocade, it has experienced the prosperity and vicissitudes of history and runs through the development of the whole Shu civilization.

第二节　蜀锦的种类和花色/Varieties and Patterns of Shu Brocade

一、蜀锦的种类/The Varieties of Shu Brocade

　　根据织物组织、生产工艺流程、使用机具，蜀锦被划分为经锦和纬锦两大类。从时间线来说，唐代中期以前均是多种色彩的经线显花，均称为经锦，其图案都是以彩色经线交织显现花纹。直到隋唐时期，特别是唐代中期后，逐渐演变为不同色彩纬线显花的纬锦，其图案都是以纬线显花，这种形式突破了经线显花中花样大小和配色的局限。宋末至明清时期，发展为既有经线显花又有纬线显花的织锦，其中不乏许多有名的蜀锦，这种形式使得纹样更为精致，能够表现的图案也更为丰富。蜀锦发展到高科技的现代，很多蜀锦已经发展成了数码织锦，这一类的织锦织造效率高，花色丰富，但是部分工艺至今仍无法在现代织机上完成。目前只有少部分的匠人使用传统织造机器织造蜀锦，这不仅是手工艺的传承，也是文化的保护。

According to the fabric weave, production process and machines, Shu brocade is divided into warp brocade and weft brocade. In terms of history, before the middle of Tang Dynasty, warp colored brocade was prevailed, and the patterns were all interwoven with colored warp lines to display patterns. Until the Sui and Tang Dynasties, especially after the mid-Tang Dynasty, it gradually evolved into weft brocade, and the patterns were all displayed by weft lines of different colors, which broke through the limitations of pattern size and color matching of warp brocade. From the late Song Dynasty to the Ming and Qing Dynasties, it developed into warp colored brocade and weft colored brocade, among which there were many famous kinds of Shu brocade. It made the patterns more delicate and could display richer patterns. In high-tech modern times, many kinds of Shu brocade are weaved digitally. Though digital-woven brocade is with high efficiency and rich colors, some traditional crafts cannot be completed on modern looms so far. At present, only a small number of craftsmen use traditional weaving machines to produce Shu brocade, which is not only the inheritance of handicrafts, but also the protection of culture.

1. 多彩经线显花的经锦蜀锦/Warp Colored Shu Brocade

经向显花蜀锦以经二重或多重平纹为组织，图案按照彩色经线起伏交织显现，显花经线会遮盖不显花经线与纬线的交织点，不同色彩的经线会表现在花样相应的色彩上。

Warp colored Shu brocade is weaved by double or multiple plain warp threads, displaying the patterns by the ups and downs of interwoven colored warp. The warp threads displaying patterns will cover the interweaving point between the warp not showing patterns and the weft, and the warp of different colors will be displayed on the corresponding colors of the pattern.

代表性的经向显花蜀锦有对龙对凤彩条经锦、水禽波纹锦、几何纹绒圈锦、长乐明光、登高明望四海、韩仁绣锦、广山锦、朱萸纹锦、太阳神鸟（山岳树木纹）、胡王锦、五星出东方锦、方格兽纹锦、对马对羊树纹锦、盘球狮象锦、球路孔雀锦、格子花纹蜀江锦等。

The representative warp colored Shu brocade include warp brocade with symmetric dragons and phoenixes patterns, waterfowl and ripple brocade, geometric-pattern loop-pile brocade, brocade with Chinese characters "CHANG LE MING GUANG (endless happiness)" (as shown in the figure), brocade with Chinese characters "DENG GAO MING WANG SI HAI (ascending high to enjoy a distant view of the whole world)", brocade with Han Ren's inscriptions, brocade with Chinese characters "GUANG SHAN (wide mountains)", cornel-pattern brocade, sunbird-pattern brocade, brocade with Chinese characters "HU WANG", brocade with Chinese characters "WU XING CHU DONG FANG LI ZHONG GUO (Five Stars out of the East Benefit China)", brocade with animal and square patterns, brocade with symmetric horses, sheep and tree leaves patterns, brocade with patterns of lion and elephant playing ball, brocade with linked pearls and peacock patterns, brocade with chickens, sheep and light trees patterns, Shujiang brocade with grid patterns, etc.

2. 多彩纬线显花的纬锦蜀锦/Weft Colored Shu Brocade

纬向显花蜀锦是指采用多色纬线、多把梭子按花纹顺序实现交替显花的织锦，突破了经线显花中花样大小及配色的限制，组织结构也从平纹、斜纹过渡到缎纹，同时，相应的生产器具、工艺及技术也有很大提高。

Weft colored Shu brocade refers to brocade with multi-color weft threads and multiple shuttles displaying patterns alternately, which breaks through the limitation of pattern size and color matching in warp brocade, and the fabric weave also transits from plain and twill to satin. Meanwhile, the production instruments, processes and technologies have also been greatly improved.

代表性的纬向显花蜀锦有花鸟纹锦、狮凤圆纹蜀江锦、鸟兽联珠纹锦、赤地花莲珠圆纹锦、联珠对雁锦、红地八角团花锦、福捧寿纹锦、墨绿地圆花锦、藻井纹彩锦、蝶纹绵、蜀香锦、菊花锦、浣溪锦等。

The representative weft colored Shu brocades include the brocade with flower and bird patterns, brocade with green land, lion and phoenix patterns, brocade with linked pearls, birds, and animals patterns, brocade of red ground with the plaid lotus patterns, brocade with linked-pearl and wild goose patterns, brocade of red ground with octagon round patterns, brocade with bat patterns and Chinese character "SHOU", brocade of green ground with flowers and plants patterns, brocade of sunk panel grains, butterfly-pattern brocade, Shuxiang brocade, chrysanthemum-pattern brocade, Huanxi brocade, etc.

3. 经纬线同时显花的蜀锦/Weft and Warp Colored Shu Brocade

经纬线同时显花的蜀锦有龟背折枝花蜀锦、八答晕锦、凤穿牡丹锦、红地万年青织金锦、双狮雪花球路纹锦、如意天花锦、云龙团花锦、龟子龙纹锦、穿花凤二龙戏珠球路锦、龙纹格子锦、红地八角团花锦、墨绿地圆花锦、福捧寿纹锦、藻井纹彩锦等。

Weft and warp colored Shu brocades include the brocade of moiré pattern and floral sprays, Badayun brocade, brocade with patterns of phoenixes in peony, gold brocade of red ground with evergreen patterns, brocade with patterns of double lions and snowflakes, brocade with patterns of Ruyi and flowers, brocade with clouds, dragon and round patterns, brocade with turtle and dragon patterns, brocade with patterns of phoenix in flowers and two dragons playing a pearl, brocade with dragon and square patterns, brocade of red ground with octagon round patterns, brocade of green ground with flowers and plants patterns, brocade of bat patterns and Chinese character "SHOU", brocade of sunk panel grains, etc.

4. 经缎地起纬浮花蜀锦/Shu Brocade with Warp Ground and Weft Patterns

极具代表性的经缎地起纬浮花蜀锦是盛于近代的"晚清三绝"的月华锦、雨丝锦和方方锦。主要特征是由经向呈现不同深浅彩条，或经向色条由细到粗再由粗到细排列具有晕裥效果，以及经纬向形成方格花纹图案。

The typical Shu brocades of this kind are Yuehua brocade, Yusi brocade and Fangfang brocade, which flourished in modern times. They are mainly characterized by forming color strips in warp direction, or arranging the color strips in warp direction from thin to thick and then from thick to thin to display the shaded effect, and forming square patterns in warp and weft directions.

二、蜀锦的主要品种/The Main Varieties of Shu Brocade

根据图案纹样与色彩在锦面的布局、结构、应用范围及工艺特征，蜀锦产品可分为方方锦、雨丝锦、月华锦、浣花锦、铺地锦、通海缎、民族缎、现代蜀锦等类。

According to the layout, structure, application scope and technological characteristics of patterns and colors on brocade surface, Shu brocade can be divided into Fangfang brocade, Yusi brocade, Yuehua brocade, Huanhua brocade, Pudi brocade, Tonghai satin, ethnic satin and modern Shu brocade.

1. 方方锦/Fangfang Brocade

方方锦就是在锦面上纵向或横向色经色纬交织或提花成条，彩条相交成"#"方格，内设图案花纹的提花织锦（图1-1）。梅兰竹菊、石榴多子、梅鹤争春、八宝八吉、莲子莲花等都是方格内常饰图案。方格兽纹经锦、联珠棋格方方锦、八宝吉祥方方锦、菱格五毒方方锦等都是典型代表。

Fangfang brocade is a jacquard brocade on which vertical or horizontal color warps and wefts interweave or form strips by jacquard weaving, and the strips intersect into "#"—shaped squares, with patterns inside (Figure 1-1). Plum, orchid, bamboo and chrysanthemum, pomegranate with many seeds, plum and crane in spring, the Eight Treasures and the Eight Auspicious Symbols, lotus flowers

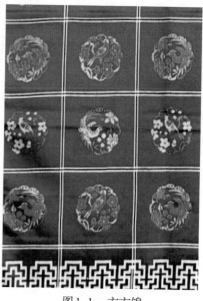

图1-1　方方锦
Fangfang Brocade

and seeds are all common patterns in squares. Warp brocade with animal grain, Fangfang brocade with linked pearls and squares, Fangfang brocade with Eight Auspicious Symbols and Fangfang brocade with patterns of diamond checks and five venoms are all typical examples.

方方蜀锦孕育于战国秦汉时期，从唐代到宋、元、明、清，蜀地方方锦随着织锦工具和技艺的不断改进和提高，在纹样色彩及织造技艺方面都不断创新，形成了独特风格。以致享有晚清蜀锦"三绝"盛誉，在这进程中织锦艺人利用各种表现技巧来充分展现蜀锦的风采。

Fangfang brocade emerged during the Warring States and the Qin–Han Period. From the Tang Dynasty to the Song, Yuan, Ming and Qing Dynasties, with the continuous improvement of brocade machines and skills, Fangfang brocade constantly innovated in pattern color and weaving skills, forming a unique style. As a result, it was renowned as one of Three Wonders of Shu brocade in the late Qing Dynasty. Brocade craftsmen used various skills to fully display its elegance.

2. 雨丝锦/Yusi Brocade

雨丝锦的锦面由白色和多色彩的经线组成，以一白一色条经丝为"雨"，"雨"内色经由多逐渐减少，白经由少渐多，按一定比例逐步过渡，形成色白相间、对比色光明亮的丝丝雨条。雨条上再饰以各种花纹图案，给人一种轻快、舒适的韵律感。雨丝锦与月华锦不同的地方是用色白经条的宽窄来达到深浅过渡的视觉效应。彩条的配色一般多用比较明快艳丽的对比色，彩条宽窄对应，既调和了对比强烈的色彩，又突出了彩条之间的花纹，达到了"烘云托月"的艺术效果。雨丝锦的品种较多，有小雨丝、大雨丝、单色雨丝和多色彩条雨丝等。因其图案内容丰富，常以各种图案花纹给产品命名，如龙凤雨丝、双龙戏珠雨丝、牡丹雨丝、仙鹤雨丝（图1-2）、文君听琴雨丝等。

The surface of Yusi brocade is composed of white and multi–color warp threads, with one white and one color strip warp threads as "rain" stripes. In the "rain" stripes, the color warp threads gradually decrease while the white ones gradually increase, and they transit according to a certain proportion, forming rain stripes with white and bright contrasting colors. The rain strips are then decorated with various patterns, presenting a lucid and lively sense of rhythm. The difference between Yusi brocade and Yuehua brocade is the application of width of white and color warp strips to achieve the visual effect of color transition. The color strips use bright contrasting colors, with different width, which not only harmonizes the contrasting colors, but also highlights the patterns. There are many varieties of Yusi brocade,

图1-2　仙鹤雨丝锦
Crane–pattern Yusi Brocade

including that of light rain, heavy rain, monochrome rain and multicolor rain, etc. Yusi brocades are often named after various patterns, such as Yusi brocade with dragon and phoenix patterns, Yusi brocade with patterns of double dragons playing a pearl, peony–pattern Yusi brocade, crane–pattern Yusi brocade（Figure 1–2）, Yusi brocade with patterns about Zhuo Wenjun and Sima Xiangru's love story, etc.

3. 月华锦/Yuehua Brocade

月华锦（图1-3）是由汉唐时期的"晕裥锦"发展而来的，也就是锦面呈多色彩条。一条完整的晕裥彩条，称为"月牙"。晕裥月华锦的工艺特点主要体现在由染色及牵经经丝排列、色筒子的增减及变换上把排列，来实施"晕裥"色彩在锦面上自然逐深或逐浅，具有中国画的晕色效果，节奏和谐犹如雨后彩虹，具有一种高超的色彩变化艺术。月华锦以"月华三闪锦""月华锦被"和"月华雨丝锦"等为代表。

Yuehua brocade（Figure 1–3）was developed from the "halo–pleated brocade" in the Han and Tang Dynasties, that is, halo–pleated brocade with multi–color strips on the brocade surface. A complete halo–pleated color strip is called a "crescent moon". The technological characteristics of halo–pleated Yuehua brocade are the dyeing, the arrangement of warp traction, the increase and decrease of color bobbins as well as the arrangement of rakes, to realize the tone–shade effects on the brocade surface. It has the halo effect of Chinese painting, like the rainbow after rain, with a superb art of color change. Yuehua brocade is represented by "Three–Flash Yuehua brocade" "Yuehua brocade quilt" and "Yuehua Yusi brocade".

提起"月华"，会使人联想到月亮的清幽光晕，因此可推断"月华"最初的艺术构想是受到了月光的启迪，而制作者的本意也正是要把他在某一瞬间从月光所获得的灵感在锦上再现，使天上的月光在人间大放异彩。当一幅月华锦展现在眼前，立即会闪出一道道美丽朦胧的光晕，若明若暗，时隐时现，给人一种"紫烟凌凤羽，红光随玉辇""方晖竟入户，圆影隙中来"的真切感受。在那无限柔和的光晕之中，点缀着各式各样的图案花纹，游鸾翔凤、比翼和鸣和奇花异卉，实在是"丝中传意绪，花里寄春情"，在不知不觉间，把人带入"照雪光偏冷，临花色转春"的意境里。

"Yuehua" in Chinese will remind people of the quiet halo of the moon. Therefore, we infer that the artistic conception of "Yuehua" was inspired by moonlight, and the original intention of the producer was to reproduce the inspiration he got from moonlight at a certain moment on the brocade. When a Yuehua brocade is displayed in front of us, it will immediately flash a beautiful and hazy halo, giving a real feeling of moonlit night. In the soft halo,

图1-3 月华锦
Yuehua Brocade

13

various kinds of patterns are decorated, such as flying phoenix, singing birds, exotic flowers and plants, which unconsciously brings people into the poetic scene.

4. 浣花锦 /Huanhua Brocade

浣花锦又称花锦，是宋代蜀锦艺人受到"流水泛波"的启发而设计的一个品种，是对落花流水锦（图1-4）的继承和发展。据说是因锦织成后，多在锦江上游的浣花溪水内洗涤而得名。浣花锦分绸地、缎地两种，纹样极为丰富，如大小方胜、梅花点、水波纹等，风格古朴、大方、典雅，是清末民国时期比较流行的一个品种。

图1-4 落花流水锦
The Brocade with Patterns of Drifting Petals and Flowing Water

Huanhua brocade, also known as Flower brocade, was designed by Shu brocade artists in the Song Dynasty inspired by flowing water, which was an inheritance and development of brocade with patterns of drifting petals and flowing water（Figure 1-4）. It is named that after brocade was woven, it was often washed in Huanhua River in the upper reaches of Jinjiang River. Huanhua brocade is a popular variety during the late Qing-Republic Period. It can be divided into silk ground and satin ground, with rich pattern designs, such as big and small square color flower, plum blossom spots and water ripples, etc. Its style is simple and elegant.

5. 铺地锦 /Pudi Brocade

铺地锦即"锦上添花锦"。缎纹组织上采用几何纹样或细小的花纹铺地，形成规矩的地面花纹，再嵌以五彩斑斓的大朵花卉，如宝相花、牡丹花等，地纹烘托主花，使其色彩更加艳丽，层次分明。有的加金线织造，极为富丽堂皇（图1-5）。

Pudi brocade is also called "brocade with patterns of more flowers". Geometric patterns or small patterns are used as ground on satin texture, and then large patterns of colorful flowers, such as Baoxiang flowers and peony flowers, are embedded. The ground patterns set off the main flowers to be more colorful and distinct. Some are woven with gold threads, extremely magnificent （Figure 1-5）.

6. 通海缎/Tonghai Satin

通海缎又称为"满花锦"或"杂花"（图1-6），锦面上的图案为多种单色或复色纹饰，且带有民族风格和地方色彩，其中百鸟朝凤、瑞草云鹤、五谷丰登、如意牡丹、龙爪菊等图案最为常见。以五枚或八枚缎纹作底，彩色纬线显花。

Tonghai satin (Figure 1–6) is also called "brocade with patterns of full flowers" or "brocade with patterns of miscellaneous flowers". The patterns on the brocade surface are various monochrome or multicolor decorations with ethic or local styles. The patterns such as birds paying homage to the phoenix, crane with auspicious plants and clouds, an abundant harvest of all crops, Ruyi and Peony, chrysanthemum, etc., are the most common. Tonghai satin is grounded with five or eight satin weaves, with multicolor weft threads displaying patterns.

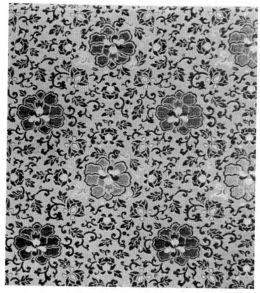

图1-5　锦上添花锦
The Brocade with Patterns of More Flowers

图1-6　通海缎
Tonghai Satin

7. 民族缎/Ethnic Satin

民族缎一般采用多色彩条嵌入金银丝织成，用作民族服饰、宗教装饰等。经纬线是纯桑蚕丝，有单色和加金线两种织造方式。其特点是锦面上的图案从经纬线交织中显现出自然光彩，富有光泽。常见的图案有团花、葵花、万字纹"卍"、寿字纹等，如图1-7所示。

Ethnic satin is generally woven with multicolored strips embedded with gold and silver threads, which is often used as ethnic costumes, religious decorations, etc. The warp and weft threads are interwoven with mulberry silk. There are two kinds of Ethnic satin, one woven with monochrome threads and the other one added with gold threads. It is characterized by the natural luster of patterns flashing from the interweaving of warp and weft. Common patterns of Ethnic satin include rounded flowers, sunflower, Chinese character "WAN" or "SHOU", "卍" symbol, as shown in Figure 1–7.

(a) 团龙民族缎/Ethnic Satin with Pattern of Winding Dragons

(b) 金丝民族缎/Ethnic Satin with Golden Threads

(c) 飞龙民族缎/Ethnic Satin with Pattern of Flying Dragons

(d) 小团龙民族缎/Ethnic Satin with Pattern of Small Winding Dragons

图 1-7　民族缎

Ethnic Satin

8. 现代蜀锦/Modern Shu Brocade

现代蜀锦，主要是指基于织锦工具、图案纹样、产品的主要组织结构的创新，开发出现代市场所需求的创新型的蜀锦产品。改革开放时期，为适应市场需求，应用传统的花楼织锦机、手拉脚踩手工木机、铁木花机、自动全铁机、剑杆织机等开发生产衣服面料、被面及装饰品。随着旅游业的发展及人们物质生活水平的不断提高，产品的需求发生了根本性的变化，急需旅游产品、纪念品、礼品及室内高档用品，诸如床上用品、坐垫、靠垫、桌旗、挂屏等产品，并要求产品色泽艳丽。图案纹样具有浓郁的地方特色与深厚的文化底蕴，强调制作精细。这种需求体现了人的价值、品质、文化艺术品位，人们渴望一种能满足精神文化需求的产品，这促使蜀锦的再次兴旺并进入一个新的发展时期。

Modern Shu brocade mainly refers to innovative Shu brocade products meeting the market demand based on the utilization of brocade tools, patterns and texture. During the period of reform and opening up, in order to meet the market demand, traditional Hualou loom, hand-pulled wood machine, iron and wooden brocade machine, automatic all-iron machine and rapier loom were used to develop and produce clothing fabrics, quilt covers and decorations. With the development of tourism and the continuous improvement of people's material living standards, the market demand has undergone fundamental changes, and there is an urgent need for tourism products, souvenirs, gifts and interior high-grade products, such as bedding, cushions, table flags, hanging panels, etc. The products are required to be rich-colored and exquisite, with strong local characteristics and profound cultural connotation. This demand reflects the improvement of people's artistic taste, which promotes Shu brocade to flourish again and enter a new era.

现代蜀锦广泛采用多色纬多梭分段换梭的织造方法，纹样上采用传统的民族图案、地方风光、几何图案和花鸟动物及著名历史字画或名人字画等，如天府锦、百子图、百宝蟠龙、百鸟朝凤、文君听琴、巴蜀览胜、蜀霞牡丹锦、熊猫嬉竹、出师表、桃园结义、太阳神鸟、芙蓉锦鸡图、九寨沟、乐山大佛、川剧脸谱、都江堰放水节、青城山、狩猎纹锦、四天王纹锦、灯笼锦、绵竹年画等。现代蜀锦，继承了传统技艺与文化，融合了时代需求而创新的织锦工艺珍品，展现了当代高超的工艺水平。

Modern Shu brocade widely adopts the weaving method of multi-color weft, multi-shuttle and segmented shuttle changing, and employs traditional ethnic patterns, patterns of local scenery, geometric shapes, flowers, birds, animals, famous calligraphy and painting, etc., such as Sichuan scenery, a hundred of boys, dragon and many treasures, birds paying homage to the phoenix, Zhuo Wenjun and Sima Xiangru's love story, a tour in Sichuan, rosy clouds and peony, pandas playing in bamboos, Zhuge Liang's poem Northern Expedition Memorial, three Chinese heroes, Sunbird, lotus and golden pheasant, Jiuzhaigou valley, Leshan Giant Buddha, Sichuan opera facial make-ups, the Water Festival in Dujiang Dam, Mount Qingcheng, hunting patterns, four heavenly kings, lantern pattern, traditional new year pictures in Mianzhu, etc. Modern Shu brocade not only inherits traditional skills and culture, but also integrates the needs of the times and innovates brocade crafts, showing the superior contemporary craft level.

三、蜀锦的花色/The Patterns of Shu Brocade

蜀锦的色彩受道教"五行"学说影响，以赤、黄、青、白、黑为五方正色，其余橙、黄、紫为间色；红灰、青灰、黄灰为复色，其余为补色。古蜀人崇拜太阳的红色，因此蜀地人们喜绮丽、鲜艳的色彩，如图1-8所示。

The color of Shu brocade is influenced by Taoist "Five Elements" theory, with red, yellow, green, white and black as the five positive colors, orange and purple as the intermediate colors; with red gray, cyan gray and yellow gray as compound colors, the rest as complementary colors. Ancient Shu people worshipped the red color of the sun, so they liked bright colors, as shown in Figure 1-8.

蜀锦配色典雅，色调浑厚，对比强烈，古朴而庄重；富有神秘浪漫色彩，具有鲜明的地方色彩和民族风格。蜀锦的配色少则两色，多则二十色；有枣红、木红、海棠红、柿红、大红、桃红、姜黄、土黄、葵黄、鹅黄、烟草棕、褐灰、烟棕、靛蓝、靛灰、淡青、菘蓝、藏蓝、秋香、墨绿、柏绿等。蜀锦的红色，历史上最为驰名，称为"蜀红锦"。

Shu brocade is elegant in color matching, sharp in color contrast, with distinct local colors and rich ethnic styles. Shu brocade has as few as two colors and as many as twenty colors, including date red, wood red, begonia red, persimmon red, scarlet, peach red, beige, yellowish brown, sunflower yellow, light yellow, tobacco brown, brown gray, indigo blue, indigo gray, light greenish

(a) 红地龙凤被面/Quilt Cover of Red Ground
with Dragon and Phoenix Patterns

(b) 百鸟朝凤被面/Quilt Cover with Patterns of Birds
Paying Homage to the Phoenix

图1-8 蜀锦的颜色
The Colors of Shu Brocade

blue, woad, purplish blue, moss green, forest green, cypress green, etc. The red color of Shu brocade is far-famed in history, and that's why Shu brocade is also called "Red Shu brocade".

蜀锦的色彩使画面层次丰富，主宾分明。"青地八达晕加金锦"（图1-9）是比较典型的蜀锦配色。用八种不同的色彩组成了晕色效果，用菘蓝、荧绿、苍绿、浅鹅黄、柿红、橘红、浅墨莲、褐棕等色织出纹样，整幅图案色调寒中带暖，色彩纷繁，繁而不乱。每组纹样色彩都是用调和色配成。整幅图案色彩配合协调，在运用色彩时，根据图案的不同结构，采用互相并列、映衬、烘托、穿插和包边等借用手法。以丰富纹样色彩的效果，使纹样更显得斑斓多彩，意蕴深沉，含蓄且耐人寻味。

图1-9 青地八达晕加金锦
The Brocade of Green Ground with
Badayun Patterns and Golden Threads

The colors of Shu brocade make the picture distinct in gradations. Brocade of green ground with badayun patterns and golden threads (Figure 1-9) uses a typical color matching of Shu brocade. Eight different colors are used to display a halo effect, and patterns are woven with woad, fluorescent green, pale green, light goose yellow, persimmon red, reddish orange, light ink lotus, reddish brown, etc. The color match of whole pattern is harmonious, complicated but not messy. Each group of patterns are matched with harmonic colors, coordinated in color match. When using colors, according to the different structures of patterns, techniques such as parallel connection, setting off, setting off by contrast, alternating and wrapping each other are adopted, so as to enrich the pattern colors.

如图1–10所示的绿地"长乐明光"锦，在墨绿地上清晰地配置了靛青、枣红、柿黄三种色彩，使文字、动物和人物色彩明艳、淡雅，再烘托以沉着的云纹，使主题更突出；在运用色彩时采用了互相借用手法，使少套色达到多套色的效果，色彩配置花地分明、色色见效，使画面更完美、更生动活泼。

图1–10　绿地"长乐明光"锦
The Brocade of Green Ground with Chinese Characters "CHANG LE MING GUANG"

As shown in Figure 1–10, brocade of green ground with Chinese characters "CHANG LE MING GUANG" is clearly equipped with three colors, including indigo, date red and persimmon yellow, on the dark green ground, which makes the colors of Chinese characters, animals and figures bright and elegant. Calm moiré patterns set off the theme more prominent. In terms of color using, the technique of borrowing each other is adopted, so that less colors can achieve the effect of multiple colors, and the color match is clear and effective, making the picture more vivid and lively.

如图1–11所示的"红地织金樗蒲罗"，在枣红地上配置泥金色，形成暖色调，它充分运用了同类色，使色彩单纯统一，由于色彩的明度与纯度较高，所以统一中又有变化，使图案色彩丰富、富丽。

图1–11　红地织金樗蒲罗
Gold Leno of Red Ground with Chupu Patterns

Gold leno of red ground with Chupu (an ancient gambling tool) patterns (Figure 1–11) matches mud gold color with the date red ground to create a warm atmosphere. It makes full use of similar colors to make color match simple and unified. Because of the high lightness and purity of colors, the patterns are rich in color, with both variation and harmony.

如图1–12所示的茶色地"花树对羊"，在茶褐色的地上配置米黄色的羊、树、蝴蝶，形

图1-12 茶色地"花树对羊"
The Brocade of Dark Brown Ground with
Patterns of Flowers, Trees and Sheep

成暖色调，充分运用了相关色相衬托，层次清晰而含蓄；图案采用深色与浅色对比，在两色交界处，淡的更觉其淡，深的更觉其深，构成一幅协调大方、幽静雅致的画面，在简单的形式中洋溢着一种朴素的美。

The brocade of dark brown ground with patterns of flowers, trees and sheep（Figure 1-12）set beige sheep, trees and butterflies on the dark brown ground to form warm colors, which makes the gradation clear and implicit by making full use of the setting off of related colors. The patterns adopt dark and light color contrast. At the junctions of the two colors, the light one is lighter and the deep one is deeper, forming a harmonious, generous, quiet and elegant picture, with a simple beauty in a plain form.

如图1-13所示的"绿地织金灯笼锦"，在烟绿色的地上配置浅绿、浅铜蓝、铜蓝、浅鹅黄、葵黄、橘红六种色彩，整体以烟绿色为主调，偏暖色，并以浅铜色点缀在橘红周围，而用葵黄勾边，使灯笼的色彩鲜明。蜜蜂则全部采用浅鹅黄色，显得更富有生命力。图案的别致处是充分利用了"活色"，各种色彩轮流调换，使每个灯笼色彩都有变化。这就把红绿不可调和的对比，巧妙地变成了调和色。烟绿色的地上陪衬橘红、葵黄色等灯笼纹样，几种颜色的配合处理得优美和谐，使整个锦面呈现出热烈欢快的气氛。

The gold brocade of green ground with lantern patterns（Figure 1-13）set six colors on the smoky green ground, including light green, light copper blue, copper blue, light goose yellow, sunflower yellow and reddish orange. The dominant shade is the smoky green. Light copper is dotted around reddish orange, while sunflower yellow is used as the edge, which makes the lantern bright in color. Bees are all of light goose yellow, which makes them more vital. It is the full use of lively colors, and the exchange of various colors that

图1-13 绿地织金灯笼锦
The Gold Brocade of Green Ground with
Lantern Patterns

make the patterns unique. It makes every lantern's colors varied and also skillfully transforms the irreconcilable contrast between red and green into harmonious color match. On the smoky green ground, the lantern patterns in reddish orange and sunflower yellow are decorated. The color match is graceful and harmonious, displaying a warm and cheerful atmosphere.

蜀锦作为一种丝织提花织锦，是蜀地历史文化的典型代表。它既具有宋锦、云锦的共性，又具有蜀锦本身的特性。在几千年历史发展进程中，蜀锦受到地域环境、历史文化、风俗习惯等因素的影响，逐步形成了独自的风格。蜀锦的风格特征呈现在纹饰、色彩等方面，而且每个朝代赋予蜀锦不同的意义。

Shu brocade, as a kind of silk jacquard brocade, is a typical representative of Shu history and culture. It not only has the commonness of Song brocade and Yun brocade, but also has its own characteristics. In the course of thousands of years of historical development, Shu brocade has gradually formed its own style under the influence of regional environment, history, culture, customs. The style features of Shu brocade are presented in pattern designs, color and other aspects, and each Dynasty endows Shu brocade with different connotations.

蜀锦的图案纹样是古蜀历史文化的重要组成部分，蜀地是道教发源地，受道教文化影响颇大，吉祥寓意是蜀地民族的传统思维，因此蜀锦传统的纹样包含精巧的构思和含蓄的寓意，体现蜀锦纹样图案的"图必有意，意必吉祥"。其图案题材内容是蜀锦艺人根据生产生活实践，从大自然、日常生活、历史故事和神话传说中取材，创造了丰富的具有浓郁民族风格的锦样。常用的图案纹样有狮、虎、鹿、象、孔雀、蝙蝠、鸳鸯、蝴蝶、鱼、蜜蜂、寿龟、云雁、仙鹤、吉羊、牡丹、莲花、芙蓉、综合变形的宝相花以及象征顺利的"寿""福"字等。此外，还根据动物、植物或文字等素材，用其形、择其义、取其音，组合成富有一定寓意和象征性的图案，成为古蜀锦纹样的重要内容。

The pattern designs of Shu brocade are an important part of the history and culture of ancient Shu. Shu is the birthplace of Taoism, greatly influenced by Taoist culture. Shu nationality advocates auspicious omens. Therefore, the traditional patterns of Shu brocade contain exquisite and implicit connotations, showing the concept that "all the patterns must be meaningful and auspicious". According to the production and life practice, Shu brocade artists draw materials from nature, daily life, historical stories and myths, and create rich brocade samples with strong ethnic style. The common patterns include lion, tiger, deer, elephant, peacock, bat, mandarin duck, butterfly, fish, bee, turtle, wild goose, crane, sheep, peony, lotus flower, hibiscus, baoxiang flower and Chinese characters of FU (happiness) and SHOU (longevity). In addition, according to the shape, literal meaning and pronunciation of animals, plants or characters, the patterns with certain connotations and symbolic meanings are combined as important pattern designs of ancient Shu brocade.

1. 战国时期的蜀锦纹样/The Patterns of Shu Brocade in the Warring States Period

战国时期的蜀锦纹样，从周代严谨、简洁、古朴的小型回纹等纹样发展到大型写实多变的几何纹（图1-14）、花草纹、吉祥如意纹，还有苗条秀丽的蟠龙凤纹，如舞人动物纹锦（图1-15）、龙凤条纹锦（图1-16）、六边形条纹锦（图1-17）、塔形纹锦（图1-18）等以几何图案为骨架，人、动物巧妙设置、紧凑、均匀、执章有序，展现了战国时期思想政治的自由以及经济文化生产的繁荣景象。

图1-14　几何纹锦
The Geometric-pattern Brocade

图1-15　舞人动物纹锦
The Brocade with Patterns of Dancers and Animals

图1-16　龙凤条纹锦
The Strip Brocade with Patterns of Dragons and Phoenixes

图1-17　六边形条纹锦
The Hexagon-patterned Strip Brocade

图1-18　塔形纹锦
The Pagoda-patterned Brocade

The patterns of Shu brocade in the Warring States Period developed from rigorous, concise and simple small fret fractal patterns in the Zhou Dynasty to concrete and changeable large geometric patterns (Figure 1-14), flowers and plants patterns, auspicious patterns and dragon and phoenix patterns. The following brocades are all good examples, such as the brocade with patterns of dancers and animals (Figure 1-15), the strip brocade with patterns of dragons and phoenixes (Figure 1-16), the hexagon-patterned strip brocade (Figure 1-17), the pagoda-patterned brocade (Figure 1-18). The above brocade varieties used geometric patterns as the framework, decorated skillfully and neatly with figure and animals, showing the freedom of social ideology and the prosperity of economic and cultural production in the Warring States Period.

2. 汉代的蜀锦纹样/The Patterns of Shu Brocade in the Han Dynasty

汉代蜀锦纹样特点为飞云流彩。从考古出土的古蜀汉锦中，有云气纹、文字纹、动植物纹等纹样，其中以山状形、涡状流动云气纹为主，这种纹饰有云气流动、连绵不绝的艺术效果。汉代的蜀锦纹样中云气纹样的不同变化，使缎面上自然形成一种典型的汉式纹样风格。

The patterns of Shu brocade in the Han Dynasty were characterized by flying clouds and flowing colors. In the ancient Shu brocade of the Han Dynasty unearthed, there are patterns of floating clouds,

Chinese characters, animals and plants, etc. Among them, mountain and flowing clouds are the main patterns, giving ethereal artistic effect. Different changes of the floating cloud patterns naturally formed a typical pattern style of Han.

祥鸟瑞兽是此时期较为流行的纹样（图1-19），以动物为题材常作侧面剪影，动物内容丰富，姿态各异，如停立、奔走、回首、争斗，形态生动，气势雄伟。"五星出东方利中国"锦（图1-20），纹样虎豹延视、狮熊嗥战、蛟龙委蛇、万兽云布，俨然展现了一幅天地和谐的画面。此外，茱萸纹（图1-21）也是汉代蜀锦使用最多的一种纹样，同时也是我国最早出现的植物纹样之一，具有佩茱萸驱灾辟邪之意，汉代蜀锦中应用的茱萸纹造型精巧，装饰优美，具有典型汉代文化的风韵。另外还有"万事如意"锦（图1-22）、"长乐明光"锦（图1-23）、"韩仁绣"锦（图1-24）等代表性品种。

Auspicious birds and beasts are popular patterns in this period (Figure 1-19), often shown as silhouettes. Animals are rich in variety, different

图1-19 对鸟对羊树纹锦
The Brocade with Patterns of Birds, Sheep and Trees

图1-20 "五星出东方利中国"锦
The Brocade with Chinese Characters "WU XING CHU DONG FANG LI ZHONG GUO"

图1-21 茱萸纹锦
The Cornel-patterned Brocade

图1-22 "万事如意"锦
The Brocade with Chinese Characters "WAN SHI RU YI"

图1-23 "长乐明光"锦
The Brocade with Chinese Characters "CHANG LE MING GUANG"

图1-24 "韩仁绣"锦
The Brocade with Chinese Characters "HAN REN XIU"

in posture, such as stopping, standing, running, looking back and fighting, vivid and imposing. Brocade with Chinese characters "WU XING CHU DONG FANG LI ZHONG GUO (Five Stars out of the East Benefit China)" (Figure 1-20) shows a harmonious picture of animals between heaven and earth, with tigers and leopards staring, lions and bears roaring, dragons winding. In addition, cornel pattern (Figure 1-21) is one of the most frequently used patterns of Shu brocade in Han Dynasty, and is also one of the earliest plant patterns in China. It is said that wearing cornel can help repel disasters and evil spirits. The Cornel patterns of Shu brocade in Han Dynasty is exquisite in shape and beautiful in decoration, with the charm of typical Han culture. Moreover, there are other representative varieties, such as brocade with Chinese characters "WAN SHI RU YI" (Good Luck)

（Figure 1-22），brocade with Chinese characters "CHANG LE MING GUANG"（Figure 1-23）and brocade with Chinese characters "HAN REN XIU"（Figure 1-24）.

不得不提的是，此时还出现一种特殊的织锦。1934年12月在蒙古塞楞河上游的诺因乌拉山坡上匈奴古墓群中出土了公元前1世纪的汉墓，其中一幅有"双禽、山岳、树木纹"图案（图1-25）的汉锦非常引人注目。不是流云走兽、茱萸铭文，而是一幅隐约可见的雄伟秀丽的自然风光。图案的中心是两座陡峭兀立的石山，山林绝壁，十分险峻。两山崖间狭长的深谷中，长有一颗高挑细长的乔木状蕨类植物，树身主干并无分枝，对称的六片大型羽状复叶从树身两侧直接伸出。另两座石山的崖间，稍宽的深谷中长有一颗珊瑚状枝干的奇树，五条粗壮的枝干似是从根部直接伸出向上生长。最中心枝条上长有三个蘑菇状冠头，中心两侧的枝条上各有两个同样形状的冠头，最外侧的两条枝条上则各长有一个冠头，这样整棵树上共长有九个蘑菇状的冠头。两座石山的山顶上各有一只前倾的凤鸟，面向这棵珊瑚状枝干的奇树。此图案正好与《山海经》中提及的"太阳神话"与"羿射十日"神话传说相切，为少见的神话题材的汉代纹饰，体现了当时人们对十日神话和太阳崇拜的观念。

(a) 细节描绘图/Detailed Drawing　　　　(b) 实物图/Physical Patterns

图1-25　双禽、山岳、树木纹锦
The Brocade with Patterns of Two Phoenixes, Mountains and Trees

During the Han Dynasty, a special brocade merits our attention. A brocade of the Han Dynasty with patterns of two phoenixes, mountains and trees (Figure 1-25), which was unearthed from the Han tombs of the first century BC in the Xiongnu cemetery on the slope of Noyon uul on the upper reaches of the Selenga River in Mongolia in December 1934, is very eye-catching. It is a picture dimly seen of magnificent natural scenery. In the center is two steep and upright rocky mountains. In the long and narrow deep valley between the two cliffs, there is a tall and slender arbor-like fern. The trunk has no branches, and six symmetrical large pinnate compound leaves extend from both sides. In a wider deep valley between the cliffs of the other two rocky mountains, there is a strange tree with coral-like branches and five thick branches seem to extend directly from the roots and grow upward. There are also nine mushroom-shaped crowns on the whole tree. On the top of each of the two rocky mountains, there is a forward-leaning phoenix. This pattern, rarely used in brocade of the Han Dynasty, echoes the "Sun Myth" and the legend of "Houyi Shooting Ten Suns" mentioned in *The Classic of Mountains and Rivers*, which reflects ancient people's worship of sun.

3. 唐代的蜀锦纹样/The Patterns of Shu Brocade in the Tang Dynasty

唐代是蜀锦发展史上最光辉的时期，这个时期的图案纹样丰富多彩，章彩绮丽，流行团窠纹与花样折枝纹，前者为"陵阳公样"，后者为"新样"。陵阳公样是益州工官窦师纶吸收波斯萨珊王朝的文化图案纹样的精华，结合本土民族文化的特点而创造的唐代风行一时的著名锦样，其特点以团窠为主题，外环围联珠纹，其团窠中央内饰对称，多隐喻吉祥、兴旺发达、雄健、权威。著名锦样有对雉、天马、斗羊、祥凤、游麟、对龙、对鹿、对鸡、对鸟、对孔雀等，纹样均齐对称，深受欢迎，流行百年之久，如图1-26所示。

(a) 联珠狩猎纹锦/The Brocade with
Linked–pearl Patterns and Hunting Patterns

(b) 花树对鹿锦/The Brocade with
Flowers, Trees and Deers

(c) 红地对马锦/The Brocade of Red
Ground with Winged Horse Patterns

(d) 狮凤圆纹锦/The Brocade with Round
Patterns and Patterns of Lions and Phoenixes

(e) 紫地唐花纹锦/The Brocade of Purple
Ground with Tang Patterns

(f) 对鹿纹锦/The Brocade with Patterns
of Double Deers

图1-26　陵阳公样
Lingyanggong Patterns

The Tang Dynasty was the most glorious period in the history of Shu brocade. During this period, the patterns were rich, gorgeous and colorful. The roundel pattern and the pattern of branches and flowers were prevailing during the Tang Dynasty. The former belonged to "Lingyanggong patterns", and the latter "New patterns". Lingyanggong patterns refered to a lot of famous brocade patterns created by Dou Shilun, an industrial official in Yizhou, by absorbing the essence of the patterns of Sassanid Empire in Persia and combining with the characteristics of local ethnic culture. Lingyanggong patterns are characterized by the roundel patterns surrounded by pearls and symmetrical interior decorations which symbolize auspiciousness, prosperity, vigor or authority. The most common patterns include pheasants, winged horses, sheep, phoenixes, Kylins, dragons, deer, chickens, birds, peacocks, etc. As shown in Figure 1-26 these patterns, popular for about one hundred years, are usually symmetrical.

"新样"是唐代开元年间益州司马皇甫所创，主要以花鸟、团花为题材，以对称的环绕和团簇形式表现，与"陵阳公样"的"团窠"截然不同，后人称之"唐花"，其题材常采用荷花、宝相花、葡萄、牡丹、芙蓉、忍冬、鸳鸯、白头翁、朱雀、鹭、鹤等。寓意吉祥的宝相团窠花鸟纹锦（图1-27），为红、绿、棕、蓝、黄五色宝相花为中心放射状团花，由中间一朵八瓣团花与外围八朵红蓝相间的小花组成团窠，团花四周百鸟争春、鸣蜂蝶飞，显现一派春意融融、生机勃勃的景象。

"New patterns", created by Sima Huangfu in Yizhou during the reign of emperor Xuanzong of Tang Dynasty, are also called "Tang patterns" later. New patterns mainly take flowers, birds, and rounded flowers as the theme, in symmetrical surrounding and cluster form, which are completely different from the round patterns of "Lingyanggong patterns".

图1-27　宝相团窠花鸟纹锦
The Brocade with Rounded Baoxiang
Flowers and Birds

Lotus, Baoxiang flower, grape, peony, hibiscus, honeysuckle, mandarin duck, pulsatilla, rosefinch, heron, crane, etc., are all common patterns. In the brocade with rounded Baoxiang flowers and birds (Figure 1-27), a five-colored (including red, green, brown, blue and yellow) Baoxiang flower in the center and eight small ones in red and blue surrounding it make up the round pattern together, and flying birds, bees, and butterflies are decorated around, displaying a scene of exuberant and lively spring.

花鸟纹在唐代中期十分流行，并生产出许多精美的产品（图1-28），如半臂背、马缟。安乐公主出嫁时蜀地进奉的"单丝碧罗笼裙"，用细如发丝的金线织成花鸟，"花卉鸟兽，皆如粟粒"，正视旁视，日中、影中各为一色。

Flower and bird patterns were very popular in the middle of Tang Dynasty. Many exquisite

(a) 赤地格子莲华纹锦 /The Brocade of Red Ground with Lotus in Grid Patterns

(b) 沉香地瑞鹿圆花绸 /Satin of Agilawood Ground with Deer in Rounded Flowers Patterns

图 1-28　唐代花鸟纹锦
The Brocads with Flower and Bird Patterns in the Tang Dynasty

products (Figure 1-28) were produced. According to the historical records, when Princess Anle got married, Shu presented a green skirt woven with gold threads as the tribute. Woven with gold thread as thin as hair, flowers and birds patterns were as exquisite and tiny as rice grains, and appeared different colors seen from the front or the side view, under the sunlight or in the shadow.

4. 宋元时期的蜀锦纹样/The Patterns of Shu Brocade in the Song and Yuan Dynasties

宋代蜀锦以冰纨绮绣冠天下，技艺之精湛、锦纹之精美，不仅继承了唐代的风格，更有了新的创新和发展。一方面发展写生图案纹样，突破了唐代对称纹样与团窠放射式纹样的固定格式；另一方面发展应用了满地规则纹样，纹样题材既保持了宋代传统，又有了新内容。

Shu brocade in the Song Dynasty was crowned with consummate crafts and exquisite patterns, which not only inherited the style of Tang Dynasty, but also had new innovation. In Shu brocade of the Song Dynasty, sketch patterns were developed, which broke through the fixed formats of symmetrical patterns and radial roundel patterns in the Tang Dynasty; moreover, regular all-over patterns were also applied. The theme of patterns not only maintained the tradition of Song Dynasty, but also had new contents.

其主流纹样一是寓意吉祥的传统纹样，如百花孔雀、瑞兽云鹤、凤穿牡丹、如意牡丹、折枝如意花纹等。二是几何图案的旋转、重叠、拼合、团叠在圆形、方形、多边几何形图案骨架中，如八答晕锦、六达晕锦，均采用牡丹、菊花、宝相花图案用虹形叠晕套色的手法，在纹样的空间镶以龟背纹连线等规则纹充满锦缎，达到了锦上添花的效果，具有特殊风格。三是在宋代锦样中还有一种特别值得称道的"紫曲水"的图案，俗称"落花流水"（图 1-29），唐诗曾有"桃花流水窅然去，别有天地非人间"以及宋词"花落水流红"等诗句，给予蜀地织锦艺人启迪与灵感，用紫曲水的纹样织造了别具一格的"落花流水锦"。其花一般以梅花或桃花为题材，体现了织锦艺人如流水般的艺术生活。"落花流水锦"未见有宋代实物传世，明代遗存十分丰富。1979 年，成都西郊明墓出土一件"落花流水锦"锦衣残片，锦料虽年久褪色，但纹样图案完整清晰，仍可见到蜀锦"紫曲水"图案纹样的特殊风貌。从这幅锦样分析，散落的梅花漂浮于水纹之上，波随风动，花随水流，情趣深厚，韵味无穷。

The mainstream patterns could be classified into three types. The first type were traditional patterns with auspicious connotations, such as flowers and peacock, auspicious animals, clouds and crane, phoenix in peony flowers, Ruyi and peony, branches and Ruyi. The second type were the rotation, overlapping and splicing of geometric patterns stacked in circular, square and multilateral

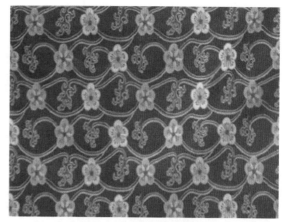

图1-29　紫曲水图案
The Brocade with Patterns of Drifting Petals and Flowing Water

geometric pattern skeletons. Badayun brocade and Liudayun brocade fell in this category, which all adopt peony, chrysanthemum and Baoxiang flower patterns, with the effect of rainbow-shaped overlapping and topping halo, inlaid with regular patterns such as tortoiseshell patterns in the full ground, adding brilliance to the brocade. The third type was a pattern named "ZI QU SHUI", also named "drifting petals and flowing water" (Figure 1-29). Shu brocade artists got inspiration from the verses of Tang poetry and Song Lyrics, and woven the brocade with the unique pattern design of purple flowing water and flower petals, usually plum or peach blossoms. Brocade with patterns of drifting petals and flowing water has not been handed down from the Song Dynasty. However, the remains of the Ming Dynasty are very rich. In 1979, a fragment of brocade clothes with this pattern was unearthed from Ming Tomb in the western suburbs of Chengdu. Although it faded over time, the pattern was still complete and clear and could be vividly seen. From the analysis of this brocade sample, the scattered plum blossoms float on the water, with profound taste and lasting appeal.

图1-30　灯笼锦
The Lantern Brocade

当时，还有一种非常流行的锦样，即灯笼锦，如图1-30所示。灯笼锦又名"庆丰年""天下乐"，一直流行到明末清初，所传不下百十种图样。此类锦样，形制奇巧，纹饰精美，称为"奇锦"。纹样中灯壁垂挂吊珠名曰"珠联璧合"；如果灯下悬坠一条玉鱼，则又成了"吉庆有余"；灯旁悬结谷穗，灯下有蜜蜂飞动，意为"五谷丰登"。以谐音和隐喻来表现人们美好的生活向往。每到传统佳节，各种花灯相继舞于市，各府第衙门亦亮灯烛，家家灯火，处处管弦，呈现出一派华灯齐放、艳彩映天的良辰美景。织锦艺人便在这种生活的启迪下，创造出了如此绚丽多彩的灯笼锦。此类纹样寓意着社会风调雨顺、国泰民安、生活美好的时代景象。

During this period, a brocade named "lantern brocade" was also popular. As shown in Figure 1-30, the lantern brocade, also known as "QING FENG NIAN (the celebration of a harvest)" and "TIAN XIA LE (a universal celebration)", was popular until the late Ming and early Qing Dynasties, and there were no fewer than 100 kinds of patterns. The lantern brocade was reputed as "strange brocade" because of its ingenious form and exquisite pattern designs. The lantern pattern was often added with some ornaments to express people's yearning for a better life with homophones and metaphors. For example, the lantern with hanging beads was called "a perfect pair", with a hanging jade fish "blessing of prosperity", with ears of grain and bees "a golden harvest". During the traditional festivals, various lanterns lightened the city, showing a splendor scene of bright lamplights and colorful glory reflecting the sky. Inspired by the festive scene, brocade artists created the brilliant lantern brocade, to symbolize the prosperity of nation and happy life of people.

元代蜀锦的织造技艺、产品风格及图案纹样仍保持着唐宋的风格，同时受到波斯的影响，产品纹样在图案上大量使用金线织造，此方式为元代织锦的一大特点，称为"纳石失""金搭子"。蜀人何绸为隋文帝仿制波斯送的织锦袍，质量优于波斯，而且蜀地金箔技艺历史悠久，细如发丝的金丝在元代广泛使用，据元人戚辅之《佩楚轩客谈》记述，蜀地主要流行的锦样有长安竹、天下乐、宝界地、八答晕、铁梗襄荷、雕团、象牙、宜男、方胜、狮团等，当时蜀地的十样锦与以往不同，都采用金线织造，锦样更加富丽堂皇，豪华富贵，如图1-31所示。

(a) 蓝地如意云纹/Patterns of Moire and Ruyi on Blue Ground

(b) 四合如意牡丹锦/The Brocade with Patterns of Ruyi and Peony

图1-31　元代锦样
Brocade Samples of the Yuan Dynasty

The weaving skills, product styles and patterns of Shu brocade in the Yuan Dynasty still maintained the style of the Tang and Song Dynasties. At the same time, influenced by Persia, a large number of gold threads were used to weave patterns, which was a major feature of Yuan brocade, named "NA SHI SHI" and "JIN DA ZI". A Shu brocade artist named He Chou once tributed a brocade robe to Emperor Wen of the Sui Dynasty, which imitated Persian weaving methods but had better quality. The gold foil technology in Shu has a long history, and the gold thread as thin as hair was widely used in the Yuan Dynasty. According to a book titled *Pei Chu Xuan Ke Tan* (*A talk in Peichu Room*) written by Qi Fuzhi, a scholar of the Yuan Dynasty, the main popular brocade patterns in Shu included Chang'an bamboo, Tianxiale, Baojiedi, Badayun, ivory, square color flower, etc., which were called "ten kinds of Shu brocade patterns". Different from the past, they were all woven with gold threads, more magnificent and luxurious, as shown in Figure 1–31.

5. 明清时期的蜀锦纹样/The Patterns of Shu Brocade During the Ming and Qing Dynasties

　　明代蜀锦继承了唐宋盛行的图案纹样，如卷草、缠枝、散花、折枝花卉纹样，并生产了许多著名的锦样，如红地万年青织金锦、双狮菊花球路锦、绿地织金灯笼锦、红地方圆格翔鹤灯笼锦、黄地/青地加金八答晕锦，并有了绿地富贵平安花缎、银白地寿字龙缎、茶青地五谷丰登寿字灯笼锦、百子图锦（图1–32）、太子绵羊锦（图1–33）、长安竹纹花缎、八宝吉祥方方锦、几何杂宝晕裥纹锦、四答晕锦，锦样颇多，足有数十种。

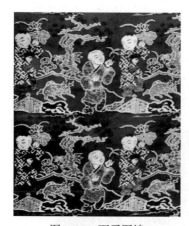

图1–32　百子图锦
The Brocade with Patterns of
a Hundred Boys

图1–33　太子绵羊锦
The Brocade with Patterns of Prince
and Sheep

Shu brocade in the Ming Dynasty inherited the popular patterns in the Tang and Song Dynasties, including patterns of rolling grass, entangled floral branches, scattering flowers and broken branches, and many famous brocade samples were produced, such as the gold brocade of red ground with evergreen patterns, brocade with patterns of balls, double lions and chrysanthemum, brocade of steel grey ground with badayun patterns, gold brocade of green ground with lantern patterns, brocade of red ground with circles, grids, cranes and lantern patterns, gold brocade of green/yellow ground with badayun patterns. Dozens of other exquisite brocades with different patterns also appeared during this period, such as satin of green ground with patterns of peonies in vase, satin of silvery white ground with patterns of dragon and Chinese character "SHOU", brocade of tea green ground with patterns of crops, lanterns and Chinese character "SHOU", brocade with patterns of a hundred boys (Figure 1–32), brocade with patterns of prince and sheep (Figure 1–33), satin with Chang'an bamboo pattern, Fangfang brocade with eight auspicious symbols, halo–pleated brocade with patterns of geometries and treasures, sidayun brocade, etc.

纹样以活泼的散花与串枝花纹图案为主，明丽中显得温柔，在质朴里显得妩媚。传统的"八仙""八宝"等图案题材也广泛应用，并成为明代织锦纹样装饰的主要素材，像其音或谐其声，并饰以随风飘舞的环带，构成一组既富于变化又和谐统一的图案。此外，新的纹样也不断发展，如灯笼锦（图1–34）、八答晕（图1–35）、天华锦（图1–36）。卷草蝴蝶纹锦的纹样以枝叶缠绕与蝴蝶变形艺术相结合，线条流畅、构思新颖巧妙、层次丰富、色调明快、图样生动活泼，长期流行。

The patterns were mainly scattered flowers and string branches, gentle, lively and simple. Traditional patterns such as "the Eight Immortals" and "the Eight Treasures" were also widely used, and had become the main materials for brocade patterns in the Ming Dynasty. These patterns were often decorated with fluttering ribbons, forming a group of patterns harmonious and full of changes. In addition, new brocade varieties with different patterns were constantly developing, such as the lantern brocade (Figure 1–34), Badayun brocade (Figure 1–35), Tianhua brocade (Figure 1–36). Brocade with patterns of rolling grass and butterflies combined the entangled floral branches with butterflies, with smooth lines, novel and ingenious conception, rich layers, bright colors and lively patterns, which had been popular for a long time during the Ming and Qing Dynasties.

图1–34　光绪灯笼锦
Guangxu Lantern Brocade

图1–35　蓝地八答晕锦
Badayun Brocade of Blue Ground

(a) 黄地小天华锦/Tianhua Brocade of Yellow Ground

(b) 宝蓝地天华锦/Tianhua Brocade of Jewelry Blue Ground

图1-36 天华锦
Tianhua Brocade

清代，特别在晚清时期，蜀锦的染织技艺已达到炉火纯青的地步，蜀锦图案纹样继承发展了唐宋以来的"富贵寿喜"传统的纹样，大量采用了如意、博古、八答晕、铺地锦、遍地方胜、八仙、八宝、八吉、龙凤、麒麟、婴戏图、写生折枝、写生团花等纹样，并生产了许多著名锦样，如八答晕锦、绕枝牡丹锦、如意天华锦、云龙团花锦、满地红孔雀羽团花锦等（图1-37）。

The dyeing and weaving skills of Shu brocade in Qing Dynasty, especially in the late Qing Dynasty, had nearly attained perfection. The brocade patterns inherited and developed the traditional patterns relevant with "wealth, longevity and blessing" since Tang and Song Dynasties. These patterns were frequently used, such as Ruyi, antiques, Badayun, more flowers, square color flower,

(a) 香色地盘绦花纹锦/The Brocade of Dark Brown Ground with Silk Ribbon Patterns

(b) 香色草纹锦/The Brocade of Dark Brown Ground with Grass Patterns

图1-37 晚清锦样
Brocade Samples in the Late Qing Dynasty

the Eight Immortals, the Eight Treasures, the Eight Auspicious Symbols, dragon and phoenix, Kylin, playing infants, floral branches and rounded flowers, etc. Many famous brocade samples were produced, such as Badayun brocade, brocade with twined branches and peony, Tianhua brocade with Ruyi patterns, brocade with patterns of moiré, dragons and rounded flowers, brocade woven from peacock feather threads of all-over red ground with rounded flower patterns, etc. (Figure 1-37).

在清代遗存的锦样中，晚清时期织制的"月华三闪"，月华锦［图1-38（a）］、雨丝锦［图1-38（b）］、方方锦［图1-38（c）］，把传统的色彩变化成色彩旋律艺术，并与创新装饰艺术结合起来，采用了多彩叠晕的佛教艺术，在丰富的色相、柔和的光晕中点缀着各式各样的图案纹样，使锦样有更加奇异华丽的效果。

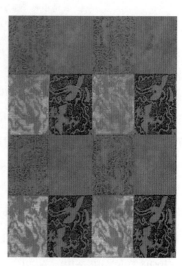

(a) 月华锦/Yuehua Brocade   (b) 雨丝锦/Yusi Brocade   (c) 方方锦/Fangfang Brocade

图1-38　月华三闪
Three-Flash Yuehua Brocade

Among the brocade samples left in the Qing Dynasty, Three-Flash Yuehua brocade, Yuehua brocade [Figure 1-38(a)], Yusi brocade [Figure 1-38(b)], Fangfang brocade [Figure 1-38(c)], were all produced in the late Qing Dynasty. They combined the traditional color strip with innovative decorative art, set various patterns in colorful ground and soft halos borrowed from the Buddhist art, fantastic and gorgeous.

6. 现代的蜀锦纹样/Modern Patterns of Shu Brocade

现代蜀锦纹样除继续采用传统图案外，还发掘创新出大量与当代社会相贴切的题材，如反应本地地域景观的都江堰（图1-39）、青城山、九寨沟、杜甫草堂、武侯祠、望江楼（图1-40）等；本地民俗，如川剧变脸、顶灯、盖碗茶、熊猫（图1-41）、芙蓉、银杏、太阳神鸟、青铜器等；使人津津乐道的故事传说，如三顾茅庐、司马相如与卓文君、薛涛吟诗等。

In addition to adopting traditional patterns, modern Shu brocade patterns are also explored and a large number of themes appropriate to contemporary society are created, such as patterns of

图 1-39 都江堰放水节
The Water Festival in Dujiang Dam

local scenic spots, like Dujiang Dam（Figure 1-39）, Mount Qingcheng, Jiuzhaigou Valley, Du Fu's thatched cottage, Zhuge Liang Memorial Hall and Wangjiang Pavilion（Figure 1-40）; patterns about local folk customs, such as face-changing of Sichuan opera, balancing a lamp on the head, covered bowl tea, panda（Figure 1-41）, hibiscus, ginkgo, sunbird, bronze ware, etc.; patterns about the well-known stories and legends, like three visits to the hut, Sima Xiangru and Zhuo Wenjun's love story, Xue Tao reciting poems, etc.

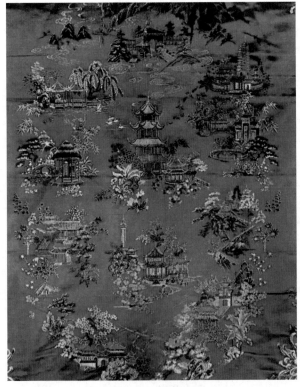

图 1-40 巴蜀胜览
A Tour in Sichuan

图 1-41 五只熊猫
Five Pandas

第三节 蜀锦的传承/The Inheritance of Shu Brocade

传统蜀锦从业人员的减少，使传统蜀锦面临消亡的压力。蜀锦传统技艺的传承，从绘稿到意匠，从拉花到投梭，主要靠艺人代代口授心传来继承，因而许多优秀产品和传统绝技不可避免地失传。传统蜀锦生产形式的转变及新型蜀锦织造工具的冲击，使人们对传统"蜀锦织造技艺"的保护越来越重视。

With the decrease of employees of traditional Shu brocade, traditional Shu brocade is facing the pressure of extinction. The inheritance of traditional Shu brocade craftsmanship, from drawing manuscripts to designing, from pattern weaving to shuttles picking, mainly depends on craftsmen's teaching in verbal form from generation to generation, so many excellent products and traditional unique skills are inevitably lost. With the transformation of traditional production form and the impact of new weaving tools, considerable attention has been paid to the protection of traditional Shu brocade craftsmanship.

为促进蜀锦传统技艺的传承和发展，国家把体现民族特色和国家水准的文化遗产和优秀的民间艺术发展为产业。2006年5月20日，国务院公布第一批国家级非物质文化遗产名录，其中"蜀锦织造技艺"榜上有名（图1-42）；2009年9月蜀锦作为"中国传统蚕桑丝织技艺"的重要组成部分入选联合国非物质文化遗产名录，得到了世界的认可与保护。

图1-42 "蜀锦织造技艺"国家级非物质文化遗产
Shu Brocade Craftsmanship as National Intangible Cultural Heritage

To inherit and develop the traditional skills of Shu brocade, China develops this cultural heritage and excellent folk art that reflects national features as industries. On May 20, 2006, the State Council announced a list of national intangible cultural heritage projects, among which "Shu brocade craftsmanship" was on the list（Figure 1-42）. In September 2009, Shu brocade, as an important part of "Sericulture and silk craftsmanship of China", was included into the Representative List of Intangible Cultural Heritage of Humanity by UNESCO, recognized and protected by the world.

"蜀锦织造技艺"凝聚了蜀地人民几千年的劳动智慧和文化精髓。为保护蜀锦，传承"蜀锦织造技艺"，成都蜀江锦院聘请蜀锦老艺人，传授年轻学工技艺，经过十多年坚守努力，现已成为蜀锦的传承基地。

"Shu brocade craftsmanship" is the crystallization of the labor wisdom and cultural essence of Shu people for thousands of years. In order to protect Shu brocade and inherit the traditional craftsmanship, Shujiang Brocade Institute in Chengdu invited old craftsmen of Shu brocade to teach the young the craftsmanship. After more than ten years of persistent efforts, it has now become the key base of Shu brocade inheritance.

为保护和发展蜀锦织造技艺，蜀江锦院近年来发掘和培养了一批蜀锦织造技艺的传承人，有国家级传承人叶永洲、刘晨晞、贺斌，省级传承人谢辉如、曹代伍，市级传承人赵兴华、李成元、季生旭、黄修忠等。此外，蜀江锦院还结合现代发展趋势，走品牌发展道路，对蜀锦产品的生产开发进行了很好的规划，纹样题材的选择、设计制作、规格和品质都有自身的特点，在消费者中树立起了"蜀江锦院"的品牌。同时不断完善"蜀江锦院"内部管理标准，规范每项工作，一步一步建立起蜀锦保护和发展的道路，如图1-43所示。成都纺织高等专科学校是四川省教育厅现代纺织技术（蜀锦）专业紧缺领域教师技艺技能传承创新平台单位，设有蜀锦织造技艺大师工作室，有助于传承、创新并保护蜀锦织造技艺。随着科技的发展，传统蜀锦织造技艺不断融入数码科技，不断创新与发展。

图1-43 蜀江锦院蜀锦织造现场
The Weaving Site of Shujiang Brocade Institute

Shujiang Brocade Institute has discovered and trained a number of inheritors in recent years, including Ye Yongzhou, Liu Chenxi and He Bin, who are national inheritors; Xie Huiru and Cao Daiwu, who are provincial inheritors; Zhao Xinghua, Li Chengyuan, Ji Shengxu and Huang

Xiuzhong, who are municipal inheritors. In addition, Shujiang Brocade Institute also follows the modern trend, takes the road of brand development, and makes an effective plan for the production and development of Shu brocade products. With the unique selection of patterns, production, specifications and quality, Shujiang Brocade Institute has built a good reputation among consumers. At the same time, it also constantly improves the internal management standards, standardizes every step, and paves the road of Shu brocade protection and growth step by step (Figure 1-43). Chengdu Textile College serves as a platform designated by the Sichuan Provincial Department of Education for inheriting and innovating Shu brocade teachers' skills. It sets a studio dedicated for Shu brocade master craftsmen, helpful for inheriting, innovating and protecting Shu Brocade craftsmanship. With the development of science and technology, traditional Shu brocade weaving skills are constantly innovating and developing by integrating with digital technology.

创新是发展蜀锦不可或缺的元素，包括图案、工艺、产品形式等方面的创新。古蜀锦纹样种类繁多、寓意深刻，有时却与现代审美格格不入，因此，创新是时代赋予蜀锦的任务。如何创新也成为现代蜀锦人面临的主要课题。所以，挖掘传统图案中的基本元素，结合现代美学及构图技巧创造新的图案，使新的构图在艺术创新的同时拥有其文化元素的支撑。要恰到好处地做到这一点并不容易，还有待进一步探求与实践。

Innovation is an indispensable element for the development of Shu brocade. The innovation of patterns, skills and products are all crucial. There are many kinds of ancient Shu brocade patterns with profound connotations, but sometimes they are incompatible with modern aesthetics. Therefore, innovation is a must for Shu brocade in modern times. How to innovate has also become the major challenge facing modern Shu brocade professionals. We should explore the elements of traditional patterns, combine modern aesthetics and composition skills to create innovative patterns, with artistic innovation and cultural elements. It is not easy to fulfill, but needs constant study and practice.

蜀锦的工艺直接影响成品的质感与品质。蜀锦的主要工艺其实已经比较完善，现有的纹织CAD辅助系统可代替并简化意匠的绘制、编制组织、储存投梭、换道、纹板等信息。但纹织CAD也只能作为提高工作效率的工具。要对工艺进行创新，除了精通纺织工艺还需大量的实践经验。织物组织和色彩的过渡都会影响成品的光泽、触感及表面呈现的纹理，如何达到理想的效果，需要设计人员的智慧和丰富的实践经验。

The technology of Shu brocade directly decides the texture and quality of finished products. In fact, the major technology of Shu brocade has been almost perfect. The Jacquard Weaving CAD System can replace and simplify the drawing and weaving of designers, store the information of picking, lane changing and pattern board. However, it can only be used as a tool to improve efficiency. In order to innovate the process, besides a good command of textile techniques, plenty of practical experience is also needed. The fabric weave and transition color will affect the luster, touch and texture of finished products. To achieve the ideal effect, it requires the wisdom and rich practical experience of designers.

蜀锦产品在形式上也有了很大的进步，近年来开发的蜀锦产品除了装饰性礼品（如卷轴、画框和摆件等），还有实用品（如名片盒、笔筒、笔袋、钱袋、靠垫、围巾等）。

Shu brocade products have been much richer in form than before. In recent years, Shu brocade products include not only decorative gifts such as scrolls, picture frames and ornaments, but also articles for daily use such as business card cases, pen containers, pencil cases, wallets, cushions, scarves and so on.

蜀锦在恢复与继承"蜀锦织造技艺"的基础上，进行了大胆的创新，依托数码电子技术，对蜀锦传统织造技术进行了改进与创新，应用数码电子技术对产品进行设计与生产（纹织CAD/CAM）加快了产品的开发速度，并提高了质量。从纹织设计、手工挑花结本到电子挑花，从传统有梭织锦机的"电子花筒"到无梭电子提花，提高了效率，现已实施了无纸版提花织绵，目前有花楼手工电子挑花织锦、电子花筒传统有梭机织锦、无梭织机电子提花机织锦、纯手工或传统有梭机械织锦等产品。

On the basis of restoring and inheriting Shu brocade craftsmanship, relying on digital electronic technology (for instance, the jacquard Weaving CAD/CAM system), the traditional weaving technology of Shu brocade has been improved and innovated. From manual pattern design and cross-stitch to electronic cross-stitch; from traditional shuttle brocade loom with the electronic cylinder to efficient shuttleless electronic jacquard machine, the design and production have been accelerated and the quality has been improved. The paperless cross-stitch and brocade weaving has been realized. At present, various innovative brocade products have been produced, including brocade woven on Hualou loom with electronic cross-stitch technique, brocade woven on the shuttle brocade loom with electronic cylinder, brocade woven on shuttleless electronic jacquard machine, all handcrafted or purely traditional brocade woven on shuttle loom, etc.

当然，目前蜀锦的产品形式还远远不够，要结合社会的发展趋势与市场走向不断开发新品种，这就需要蜀锦人孜孜不倦地努力与追求。

Generally speaking, the current product forms of Shu brocade are far from enough. It is necessary to develop new varieties in combination with the social trend and market demand, which calls for persistent efforts of all the Shu brocade professionals.

○ 第二章

宋锦
Song Brocade

一、宋锦的起源/The Origin of Song Brocade

北宋建制后，结束了五代十国的割据分裂局面，人民休养生息，社会经济得以迅速恢复和发展，封建朝廷每年官用绢帛的数量比唐朝更多。一方面，因为绢帛是宋代对辽、西夏屈服所输岁币、向金纳贡以及对外贸易的主要物资；另一方面，宋时不但沿袭了前朝的统治机构，并增设了许多官位，只要身入仕途，还另给绫绢罗锦。按宋代制度规定，每年需按品级分送"臣僚袄子锦"给所有的文武百官，其花纹各有定制，有翠毛、宜男、云雁、瑞草、狮子、练雀、宝照（有大花锦和中花锦之分），共计七等。为适应和满足朝廷对绢帛和各式织锦的特殊需要，宋代少府监下辖有绫锦院、绣局、锦院等，规模都很可观。同时在成都设有转运司、茶马司、锦院，由监官专门监制织造西北方和西南少数民族喜爱的各式花锦。

After the establishment of the Northern Song Dynasty, the separatist regimes of the Five Dynasties and Ten Kingdoms came to an end. People recuperated and recovered from the effects of war, and the social economy also recovered and developed rapidly. The amount of silk used by the royal court every year was more than that of the Tang Dynasty. On the one hand, silk were the main materials tributed to the Liao Dynasty, the Western Xia regime and the Jin Dynasty by the court of Song Dynasty and the major products of foreign trade. On the other hand, the Song Dynasty not only followed the ruling institutions of the previous dynasty, but also added many official positions. If officials assumed office, they were presented silk and brocade. According to the system of the Song Dynasty, brocade for official court dresses must be distributed to all civil and military officials according to their official ranks every year, and the patterns were customized, including green feather, daylily, wild goose, auspicious grass, lion, terpsiphone paradisi, Baozhao (divided into big flower brocade and middle flower brocade), seven grades totally. In order to adapt to and meet the special needs of the imperial court for silk and various brocade, Shao Fu Jian (a governmental agency in charge of royal clothes, treasures and food) in the Song Dynasty, set up institutions respectively responsible for silk, embroidery, brocade, etc., all in considerable scale. Meanwhile, the transshipment department, the tea and horse department and the brocade institution were also set up in Chengdu, where a supervisor specialized in weaving brocades needed by ethnic minorities in northwest and southwest.

尤其是宋高宗南渡后，北方大批统治阶级和官商巨室以及农民、手工业者都纷纷南迁，来自北方的居民"竟数倍于土著"。这样，市场上的丝织品销路激增，大大刺激了南方的丝织业生产，苏州、杭州等地涌现了大批专业性较强的手工业作坊，拥有大批掌握专门技能的工匠，能生产出大批闻名全国的各色织锦。

Especially after Emperor Gaozong of the Song Dynasty transferred the capital to Lin'an (today's Hangzhou), a large number of officials, businessmen, farmers and craftsmen in the north moved the south, and the residents from the north were several times as many as local people. In this way, the sales of silk products in the market surged, which greatly stimulated the silk production in the south. A large number of handicraft workshops emerged in Suzhou and Hangzhou, with a great many craftsmen with specialized skills, who could produce various brocades nationally renowned.

宋锦就在这样的时代背景下应运而生。

Song brocade came into being under such an historical background.

二、宋锦的发展/The Development of Song Brocade

"上有天堂，下有苏杭"这句话是在宋代开始流行的，那时的长江三角洲丰饶得就像一个巨大而殷实的粮仓，故有"苏湖熟，天下足"的民谚。"吴中一年蚕四五熟，勤于纺绩"，且技术水平也趋于全国领先地位。苏州地区很快出现了"丝绵布帛之饶，覆衣天下"的盛况。著名的苏州宋锦就从这一时期逐步兴起。

As a popular Chinese saying goes, "there is a paradise in heaven, while there are Suzhou and Hangzhou on earth." The saying became popular just in the Song Dynasty. During that time, the Yangtze River Delta was as rich as a huge and abundant granary, as the folk proverb goes, "When Suzhou pefecture and Huzhou perfecture have a bumper harvest, the entire country has enough food." According to the historical records, "In Wuzhong silkworms ripe four or five times a year, so people here are diligent in spinning and weaving." Their technical level also held the lead in the country. Silk and brocade produced in Suzhou were sold and used nationally. The Song brocade in Suzhou gradually prospered during this period.

苏州地处太湖之滨，千里沃野，遍地蚕桑，历来为锦绣之乡，是我国著名的丝绸古城。图2-1~图2-3呈现了古城苏州的生活场景。

Suzhou is located on the shore of Taihu Lake, with thousands of miles of fertile land and

图2-1　古城苏州
The Ancient City of Suzhou

图2-2　遍地蚕桑的苏州
Suzhou with Sericulture Everywhere

mulberry silkworms everywhere. It has always been the hometown of silk and brocade in China. Figures 2-1 to 2-3 show the scenes of life in the ancient city of Suzhou.

　　在吴地人的心目中，没有一样东西像蚕这样宝贝、高贵、神圣，因此，人们亲昵地称它为"蚕宝宝"；将即将上山的蚕看作即将出嫁的闺女，亲热地称为"蚕花娘娘"；还建造寺庙，如苏州盛泽镇上的蚕王殿，用以祭祀供奉"蚕花娘娘"。在每年旧历三四月便是"蚕月"，是一年中最忙的季节，又是采桑，又是饲养，还要贴门神护蚕，对蚕是百般呵护。

图2-3　洗蚕匾
Washing the Silkworm Rearing Trays

In the minds of Wu people, nothing is as precious as silkworms. They call them "silkworm babies" affectionately, and call the silkworm about to make cocoons "Silkworm Empress". They even built temples, such as the Silkworm King Hall in Shengze Town in Suzhou, to worship the "Silkworm Empress". The third and fourth lunar months are the "Silkworm Months", the busiest season of the year. Wu people picked mulberry leaves and bred silkworms. They often sticked a door-god on doors to protect silkworms.

　　然而，农家养蚕的生活又是辛劳和艰难的，即使农家全力投入蚕业生产，也并未带来幸福生活，因为，一番劳作为的是交纳统治者的苛捐杂税。

The sericulture was hard and difficult. Sometimes, even if farmers devoted themselves to sericulture production, they could not enjoy a happy life because of exorbitant taxes and levies of rulers.

　　旧时对于种桑真可说是精心备至，其工序按时令进行操作，非常细致。大体是：正月，立春、雨水，天晴时种桑秧、修桑；阴雨时，撒蚕沙、编蚕帘，蚕簹，桑剪。二月，惊蛰、春分，天晴时浇桑秧，阴雨时修桑，捆桑绳，接桑树。三月，清明、谷雨，天晴时浇桑秧，阴雨时把桑绳，修桑具、丝车。四月，立夏、小满，天晴时，谢桑、压桑秧、浇桑秧、剪桑，雨后还要看地沟桑秧，还要买粪桑……一直到十月，还修桑、把桑，忙个不停。

In the old days, mulberry planting was carefully prepared, and its working procedure was operated on time, which was very meticulous and tedious throughout all the 24 solar terms. In the first month of the lunar calendar, when it was fine, farmers planted mulberry seedlings and trimmed mulberry trees; when it rained, they scattered silkworm excrement, weaved silkworm curtains, etc. In the second month of the lunar calendar, when it was fine, they watered mulberry seedlings; when it rained, they pruned mulberry trees, tied up mulberry ropes and grafted mulberry seedlings. In the third month of the lunar calendar, when it was fine, they irrigated mulberry seedlings; when it rained, they repaired mulberry tools and silk machines. In the fourth month of the lunar calendar, when it was fine, they manured, layered, watered and trimmed mulberry seedlings; when it rained, they attended to the mulberry seedlings after the rain and buy manure, etc. Until October of the lunar calendar, they were still busy trimming mulberry trees and making mulberry ropes.

蚕种培育、改良和优化，是蚕桑养殖赖以生存与发展的基础。谈到为近代苏州蚕桑养殖的发展所作的贡献，首推浒关蚕种场，可谓蚕桑霸主，既是江苏最大的，也是我国建场历史最悠久的蚕种场之一。浒关蚕种场培养的蚕种，不仅提供给全国各地的用户，如浙江、福建、新疆等地，还走出国门，供应给阿尔巴尼亚、越南、朝鲜等国，在他国异乡生根发芽、开花结果，甚至提供技术指导，开办蚕种场。

The rearing, improvement and optimization of silkworm are the basis for the survival and development of sericulture. Xuguan Silkworm Farm is not only the largest, but also one of the oldest silkworm farms in China, and it has made much contribution to the development of sericulture in Suzhou. The silkworm cultivated in Xuguan Silkworm Farm are not only provided to users all over the country, such as Zhejiang, Fujian, Xinjiang and other provinces and cities, but also supplied to Albania, Vietnam, Democratic People's Republic of Korea and other countries. The farm even provides technical guidance for those who will operate silkworm farms.

这里值得一提的是蚕桑学家郑辟疆先生，他是苏州蚕桑业科技兴起与发展的先驱。1918年，在他任浒墅关蚕校校长之职后，即引进先进的技术，对传统蚕丝业进行改革，并在1926年与蚕业专家邵申培一起集资创办大有蚕种场，即江苏省浒关蚕种场的前身。因当年为丙寅年，又因浒地原名为虎，故以"虎"为蚕种的注册商标。蚕种培育和养蚕技术的创新，使蚕农经济效益大为改观，更使中国的蚕桑及丝绸业迎来了有史以来的大发展。中华人民共和国成立前，浒关有大大小小23个私营蚕种场，之后，浒关各家蚕种场先后进行"对私改造"，变私营为公私合营，再过渡到国有经营；1956年，合并成五个国营场，其中包括一个原种场，四个普种场。产量逐年增多，桑田近200亩，生产办公用房多达3000间，平日基本员工有500人左右，最忙时，所用季节工则要超过3000人，可见当时的育种养蚕确实十分壮观。

Who is worth mentioning here is Mr. Zheng Pijiang, a sericulture scientist as well as a pioneer in the rise and development of sericulture science and technology in Suzhou. In 1918, after he became the principal of Xushuguan Silkworm School, he introduced advanced technologies to reform the traditional silk industry. In 1926, he raised funds with Shao Shenpei, a sericulture expert, to establish Dayou Silkworm Farm, which was the predecessor of Xuguan Silkworm Farm. Because that year was the year of Bingyin, and the original name of Xuguan was "HU", pronounced similarly like "Tiger" in Chinese, so they registered "Hu(tiger)" as the trademark. The innovation of silkworm cultivation and sericulture technology had greatly improved the economic benefits of farmers, and made China's sericulture and silk industry usher in an enormous growth in history. Before the founding of the People's Republic of China in 1949, there were 23 private silkworm farms in Xuguan. After 1949, each silkworm farm successively carried out the policy of "the socialist transformation of private industry", transformed into public–private partnership first and then to state–owned operation. In 1956, they merged into five state–owned farms, including one for original silkworms and four for general silkworms. The output increased year by year. There were nearly 200 mu(a unit of area, about 0.066,7 hectares)mulberry field, as many as 3,000 production and office buildings, about 500 basic employees on weekdays and more than 3,000 seasonal workers in the busiest time. All the data above indicate the magnificent scene of silkworm breeding.

明清时期是宋锦的黄金时代。公元1368年，明初洪武元年开设的苏州织造局，位于苏州市区天心桥东面，有房屋245间，织机173张，额定岁造上用锦缎1534匹，更多的则承担着临时差派任务，如明朝天顺四年（1460年），苏、松、杭、嘉、湖五府，就在常额外增造彩锦7000匹。弘治十六年（1503年），苏杭两局曾增织上贡锦绮24000匹，如此等等的临时任务常常会令额定数量高出几十倍。在明代中叶，朝廷除官府、宫匠织造外，不得不再雇用民间机户包揽领织，才能完成织造任务。故这时的宋锦业官办、民办产销两旺，各地商贾云集，盛极一时。成化、弘治年间（1465～1505年）为繁盛时期。明代有"吴中多重锦，闽织不逮"之称。

The Ming and Qing Dynasties were the golden age of Song brocade. Suzhou Weaving Bureau, located in the east of Tianxin Bridge in Suzhou City, was established in 1368. It had 245 houses and 173 looms, with a rated output of 1,534 bolts of brocade for the royal court and more temporary assignments. For example, in 1460, the five districts—Suzhou, Songjiang, Hangzhou, Jiaxing and Huzhou, undertook the extra task of 7,000 bolts of brocade. In 1503 AD, the Suzhou and Hangzhou bureaus shouldered the task of 24,000 bolts of brocade as tribute. Extra duties like this often made the output much higher. During the middle of the Ming Dynasty, besides the government and palace employed craftsmen, the imperial court had to hire more folk workers so as to complete the weaving task. Therefore, during the Ming Dynasty, the production and sales of the government–run and private Song brocade workshops were booming, and merchants from all over the country gathered and flourished. During the reign of Ming Emperors Chenghua and Hongzhi(1465–1505), the Song brocade industry reached the height of its prosperity. In the Ming Dynasty, it was said that "the brocade in Wuzhong is very exquisite, and that of Min(roughly in today's Fujian Province)is no match for it."

明清两代有很多宋锦精品传世，如宣德年间精巧绝伦的重锦《昼锦堂记》，万历年间完成的《大藏经》的裱面及织有熠熠发光的真金线的"盘绦花卉锦"，以及现分别收藏于故宫博物院和苏州博物馆的王文肃公夫妇合葬墓和元代的曹氏墓出土的宋锦文物等。另外，嘉靖末年（1566年）从严嵩的抄家物资中，有各色宋锦达87匹，也是宋锦中的精品。嘉靖年间《吴邑志》记载，明中叶时苏州有"东北半城，万户机声"之称，形容苏州东北半城专业丝绸生产区域的景观。

There are many fine works of Song brocade handed down from Ming and Qing Dynasties. For example, the exquisite brocade "ZHOU JIN TANG JI (*The Story of Zhou Jin Hall*, an article written by Ouyang Xiu, a famous scholar of Song dynasty)" woven during the reign of Emperor Xuande, the brocade book cover of *The Great Tripitaka* during the reign of Emperor Wanli, the gold brocade with silk ribbon patterns, the Song brocade cultural relics unearthed from the tomb of Wang Xijue (the cabinet minister of Ming Dynasty) and his wife as well as the tomb of Lady Cao now preserved in Suzhou Museum and the Palace Museum now. In addition, in the last year of the reign of Ming Emperor Jiajing (1566), 87 bolts of Song brocade, which were all the fine works, were searched and confiscated from the house of Yan Song, the cabinet minister of Ming Dynasty. According to *The Local Chronicles of Wuyi*, during the middle of Ming Dynasty, Suzhou was described as "the sound of thousands of looms can be heard in the northeast half city", which shows the prosperous scene of silk production in the northeast half city of Suzhou.

清初宋锦织物的作用范围扩大，清代苏州织造局的产量、规模均为"江南三织造"之首。康熙乾隆年间（1662～1795年），苏州出现了宋锦历史上的全盛时期。坐在织机的花楼上面的牵花工口唱手拉，按挑花纹样提综，坐在下面机坑潭里的织工闻歌默契配合，每提一次综，就织入一根纬线，织机发出"咿咿呀呀"的声音，织入千千万万根的纬线，就形成了一匹匹的织锦。延续到清代的乾隆盛世，苏州仍然"郡城之东皆司机业"。除了织造丝绸锦缎，打综掏泛、插丝调经、牵经接头、挑花结本等众多辅助行业也在东北半城盘结。当时苏州有十几万人从事与丝织相关的行业。"东北半城，万户机声"的盛况，造就了苏州的高度繁荣，引来了乾隆皇帝的多次南巡。

At the beginning of Qing Dynasty, the usage of Song brocade became wider and wider. The output and production scale of Suzhou Weaving Bureau in Qing Dynasty ranked first in three Silk Factories in regions south of the Yangtze River. During the reign of Emperor Kangxi and Emperor Qianlong (1662–1795), Song brocade had its heyday in Suzhou. The drawer sitting on the Hualou loom chanted and pulled, lifting the heald according to the patterns, while the weaver sitting in the pit under the loom cooperated well with the drawer. Every time the heald was lifted, a weft thread was woven. Thousands of weft threads were woven in the constant squeaking Hualou loom, forming bolts of exquisite brocades. Until the prosperous reign of Qing Emperor Qianlong, brocade workshops were widely distributed in the eastern Suzhou city. In addition to silk and brocade weaving industry, many auxiliary industries also gathered there. More than 100,000 people in Suzhou were engaged in silk-related industries. The grand occasion of brocade production had created a high prosperity in Suzhou and attracted Emperor Qianlong to make several inspection tours to the Southern China.

公元1759年，乾隆二十四年，苏州画家徐扬绘制了长达12.55m的巨幅纪实图画——《姑苏繁华图》，又名《盛世滋生图》，以恢宏的气势，形象地反映了苏州城市从阊门，经山塘，到木渎一线的商贾云集、商品林立、贸易繁盛的景象。图2-4画面中"阊门内外，居货山积，行人流水，列肆招牌，灿若云锦"。仅标出有市招的店铺就达230多家，共50多个行业，而最引人注目的是丝绸业的店铺与行会，多达14家，其中最大的一家有七间门面，还有一家二层楼五间门面，可以看出当时苏州丝绸织锦业的繁荣。

图2-4 《盛世滋生图》(局部)
Flourishing City of Suzhou(Partial)

In the 24th year of reign of Emperor Qianlong（1759）, Suzhou painter Xu Yang drew a huge realistic painting with a length of 12.55 meters titled *Flourishing City of Suzhou*, which vividly depicted the prosperous scene of Suzhou city from Changmen, Shantang to Mudu. As shown in Figure 2-4, inside and outside Changmen were full of numerous shops, crowded with pedestrians. There were more than 230 shops with signs of more than 50 industries. Among them the most striking ones were the shops and guilds of the silk industry, with as many as 14. The largest shop had seven door facades and another one a two-floor building with five door facades.

据《苏州府志》记载，当时生产织造的品种花色还有海马、云鹤、方胜、宝相花等，均是以宋代和明代留传下来的纹样仿制生产的，故又称仿古宋锦。宋代早期的图案年久失传，直至康熙年间，始有人从泰兴季氏处购得宋裱《淳化阁帖》十帙，揭取其上所裱宋锦二十种，转售给苏州宋锦机业，使早期失传的宋锦珍品得以重新组织生产，并加以创新改良，使其胜于原貌。

According to *The Records of Suzhou Prefecture*, the varieties and colors produced at that time still included seahorses, cranes, square color flower, Baoxiang flowers, etc. Since the patterns above were all copied from that handed down from the Song and Ming Dynasties, they were also called antique Song brocade. The patterns in the early Song Dynasty were lost for a long time. Until the reign of Emperor Kangxi period, someone bought *Chunhua Getie* (a book containing models of various genres of Chinese calligraphy) from Ji family in Taixing, peeled off 20 Song brocade decorated on it, and sold them to a silk shop in Suzhou as samples, so that the lost Song brocade could be reproduced. The later Song brocade handicraftsmen even made much innovations and created more novel patterns.

故宫博物院收藏的《西方极乐世界图轴》是乾隆时期苏州织造局织造的宋锦中的重锦：448cm高、196.5cm宽的独幅纹样中有278个神态各异的人物佛像，宫殿巍峨，宝池树石，祥云缭绕，奇花异鸟，它的结构之复杂，图案之精美，色彩之丰富，工艺之精湛，堪称稀世珍宝（详解于后）。另外，如云地宝相花纹重锦及加金缠枝花卉天华锦等，均为苏州宋锦中的精品。

The Sukhavati Brocade Scroll woven by Suzhou Weaving Bureau during the reign of Emperor Qianlong, a classic among the Palace Museum collection of silk and embroidery, belongs to the double brocade of Song brocade. In a single brocade with a high of 448 centimeters and a width of 196.5 centimeters, there were 278 Buddhist figures with different gestures and facial expressions, towering palaces, trees, stones, clouds, flowers and birds. Its complex composition, exquisite patterns, rich colors and superb craftsmanship made it a rare treasure (explained in detail in next section). Moreover, the double brocade of moiré ground and Baoxiang flower patterns and the Tianhua brocade woven with gold threads with patterns of entangled floral branches are also fine works of Song brocade produced in Suzhou.

苏州吴江太湖沿岸的乡民皆以饲蚕为主要副业，农家将自育之鲜蚕茧缫制成生丝，或自织，或出售。数千年不断地传承总结，加以提高，使苏州地区所生产的生丝成为中国质量最佳的生丝，其中以明代异军突起的辑里丝为代表，其后又有香山丝问世。

Most villagers along Taihu Lake in Wujiang of Suzhou took silkworm breeding as their main sideline. They reeled the self-produced fresh cocoons into raw silk, and then wove them by themselves or sold them. After thousands of years of continuous inheritance and improvement, the raw silk produced in Suzhou has become the best quality raw silk in China, which is represented by Jili silk of the Ming Dynasty, and then the later Xiangshan silk.

19世纪60年代后，吴江震泽一带的丝商和蚕农为适应出口以符合当时国外丝绸机械织造工艺之需，将辑里丝再加工，即经过拍松，剔除糙粗，再加重摇，绕一周长1.5m，分成粗、中、细三个档次，定名为辑里干经，鉴于出口为主，又名洋经。辑里丝经可使国外丝织厂商减少一道络丝工序，颇受欢迎，欧美各国厂商称之为复摆丝，在法国里昂丝市上售价每公斤63法郎，而普通白丝仅47法郎。辑里丝"颜色纯白，光泽艳丽，质地坚韧，弹性丰富，条份匀整，均非世界各生丝所可比拟"。在国内专供官府，用于织造宫廷丝织品，在近代还大量出口，享誉海内外市场。

After the 1860s, in order to adapt to the export and meet the needs of foreign silk mechanical weaving technology, silk merchants and sericulture farmers in Zhenze Town of Wujiang reprocessed Jili silk through fluffing up, removing rough silk and rereeling. The reprocessed silk, named Jili Dry Warp and sometimes called as Foreign Warp, could be divided into three grades—coarse, medium and fine. Jili Silk Warp helped reduce one reeling process for foreign silk weavers, so it was very popular among manufacturers in Europe and America. It was sold for 63 francs per kilogram in Lyons, France, while ordinary white silk was only worth 47 francs. According to the historical records, Jili silk "is pure white in color, bright in luster, tough in texture, high in elasticity and even in size, which are unmatched by other raw silk in the world." In China, it was specially used for the government to weave court silk fabrics, and was also exported in large quantities in modern times, enjoying a good reputation in domestic and overseas markets.

清光绪年间成立了经业公所，清末民初丝经行业成为一个新兴行业，吴江西南操持此业者人众10万，且扩展到苏州，集中于官太尉桥一带，为苏城纱缎业供应经纬原料。

During the reign of Emperor Guangxu in the Qing Dynasty, the Silk Warp Industry Office was established, and the silk warp industry became a new industry in the late Qing Dynasty and early Republic of China. There were about 100,000 people engaged in this industry in the southwest of Wujiang. The workers also expanded in Suzhou, who gathered in the area of Guantaiwei Bridge, suppling raw silk warp and weft for the yarn and satin industry in Suzhou.

苏州蚕桑、制丝业的发达，为苏州丝织业，尤其是宋锦业的繁荣打下了基础，提供了优质的原料保证。

The development of sericulture and silk industry in Suzhou laid a foundation for the prosperity of local silk weaving, especially Song brocade, and provided high–quality raw materials.

第二节　宋锦的种类与花色/Varieties and Patterns of Song Brocade

一、宋锦的种类/The Varieties of Song Brocade

宋锦根据其结构的变化、工艺的精粗、用料的优劣、织物的厚薄以及使用性能等方面，分为重锦、细锦、匣锦和小锦四类，也可以将重锦、细锦归纳为大锦，即大锦、匣锦和小锦三类，它们各有不同的风格和用途。

Song brocade can fall into four categories with different textures, processes, raw materials, fabric thickness and functions: double brocade, fine brocade, box brocade and small brocade.

Double brocade and fine brocade can also fall into the category named big brocade. These brocade varieties have different styles and uses.

1. 重锦/Double Brocade

重锦是宋锦中最贵重的品种，它常以精练染色的蚕丝和捻金线或片金为纬线，在三枚经斜纹的地上起各色纬花。其金线一般用以装饰主花或花纹的包边线，并采用多股丝线合股的长抛梭、短抛梭和局部特抛梭在花纹的主要部位作点缀。重锦的质地厚重精致，花色层次丰富、造型多变、绚丽多彩，产品主要是宫廷、殿堂、室内的各类陈设品，如各类挂轴、壁毯、卷轴等，如《西方极乐世界图轴》（图2-5）。

图2-5 《西方极乐世界图轴》
The Sukhavati Brocade Scroll

Double brocade is the most valuable variety in Song brocade. What is special about Double brocade is that the wefts use refined dyed silk, twisted or flaked gold threads, and that silk is combined with threads by applying the techniques of long-range shuttle, short-range shuttle and special shuttle for particular parts. Double brocade, with thick and delicate texture, rich colors and abundant patterns, is often used as furnishings in royal palaces, halls and rooms, such as hanging scrolls, tapestries and scrolls. *The Sukhavati Brocade Scroll* (Figure 2-5) is one of these.

2. 细锦/Fine Brocade

细锦是宋锦中最基本、最常见、最有代表性的一种。细锦的风格、组织和工艺与重锦大致相近。只是所用的丝线较细，长梭重数较少，地经与面经的配置比例和组织多有变化，并常以短抛梭织主体花，以长抛梭织几何纹及花的枝、叶、茎和花纹的色包边线等。以其中一组或两组短抛梭来变换色彩，不增加其厚度。原料除有全桑蚕丝外，近代多采用桑蚕丝与人造丝交织，以降低成本。故细锦易于生产，厚薄适中，广泛用于服饰、高档书画及贵重礼品的装饰装帧等。细锦图案一般以几何纹为骨架，内填以花卉、八宝、八仙、八吉祥、瑞草等纹样，典型品种有盘绦花卉锦、菱格四合如意锦（图2-6）等。

图2-6 菱格四合如意锦
Song Brocade with Patterns of Diamond Checks and Ruyi

Fine brocade is the most common and representative variety of Song brocade. Its style, texture and production processing are roughly similar to that of double brocade. However, its silk thread is thinner, the long shuttles used are less, and the proportion and fabric weave of bottom warp and surface warp are more diversified; the main pattern is often woven by short–range shuttles, and the geometric patterns, the patterns of branches, leaves and stems, the colored edge threads are often woven by long–range shuttles; one or two sets of short–range shuttles are used to change colors without increasing the thickness. Besides using mulberry silk totally, mulberry silk and rayon were often interwoven in modern times to reduce costs. As its thickness is well suited and it's easy to produce, fine brocade is widely used in attires, the mounting of top–quality paintings and calligraphy, as well as the decoration and wrapping of valuable presents. Fine brocade generally takes geometric patterns as the skeleton, filled with patterns of flowers, Eight Treasures, Eight Immortals, Eight Auspicious Symbols, auspicious grass, etc. The typical subtypes include Song brocade with patterns of flowers and colorful silk ribbons and Song brocade with patterns of diamond checks and Ruyi（Figure 2–6）.

3. 匣锦/Box Brocade

匣锦（图2-7）是宋锦中变化出的一种中档产品，它采用桑蚕丝、棉纱和真丝色绒（真丝色绒是不加捻或加少量捻的精练染色的蚕丝）交织的工艺。花纹图案大多为满地几何纹或自然型小花，以对称、横条形排列为主，色彩对比强烈，风格粗犷别致。织造时多数采用一两把长抛梭织地纹和花纹，再加一把短抛梭点缀。质地较疏松，织成后常在背面涂一层薄浆，使之挺括。一般用作中低档的书画、锦匣、屏条等的装裱。

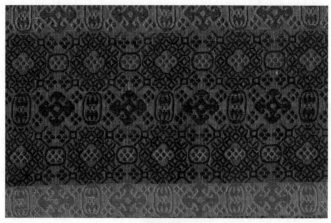

图2-7　匣锦
Box Brocade

Box brocade（Figure 2–7）is a medium–quality variety of Song brocade. It adopts the technique of interweaving mulberry silk, cotton yarn and colored silk yarn（the refined dyed colored silk without twisting or with slight twisting）. The pattern designs on box brocade center on geometrical patterns or small flowers in symmetrical or linear arrangement, the color contrast is strong, and the style is bold and unique. One or two long-range shuttles are used to weave ground patterns and main patterns, and then a short-range shuttle to embellish. Its texture is soft and loose, so after weaving, it is starched on the back for stiffness. Song brocade is often applied to the mounting of paintings and calligraphy, boxes and hanging scrolls, etc.

4. 小锦/Small Brocade

小锦（图2-8）是宋锦中派生出的又一种中低档产品，实际上它不应属于宋锦，但因其与宋锦一样也是作装裱之用，且与宋锦在同一工厂生产，故将它归入广义的宋锦大类中。小锦多数为平素或单层小提花织物，采用彩色精练蚕丝作经线，生丝作纬线，配置不同色彩和花纹，形成风格各异的织物，如彩条锦、月华锦、万字锦和水浪锦等。小锦质地轻薄，成品需用传统的石元宝进行砑光整理。小锦适用于装裱精巧的小型工艺品锦匣，如扇盒、彩蛋匣、银器匣镶边等。

Small brocade (Figure 2-8) is another low-and-medium-grade product derived from Song brocade. Strictly speaking, it should not be classified into the category of Song brocade. However, because it is also used for mounting and produced in the same workshops, it

图2-8　彩条小锦
Small Brocade with Color Strips

is also classified into the broad category of Song brocade. Small brocade is mostly plain or single-layer small jacquard fabric, which uses colored refined silk as warp and raw silk as weft. With different colors and patterns, fabrics with different styles are formed, such as the small brocade with colored strips, Yuehua brocade, brocade with Swastika symbols and wave-patterned brocade. The finished products of small brocade need to be calendered with the stone ingot, a traditional calendering tool. Small brocade is light and thin, thus suitable for decorating small objects and making brocade boxes, such as fan boxes, egg boxes or silverware boxes.

二、宋锦的花色/The Patterns of Song Brocade

（一）早期宋锦的艺术风格/The Artistic Style of Early Song Brocade

对从新疆阿拉尔出土的北宋锦袍进行分析可知，其袍料有球路双鸟纹锦、球路双羊回纹锦（图2-9）、灵鹫对羊纹锦、重莲团花锦（图2-10）等。其锦纹上的球路对鸟、双羊、双兽等均为唐代流行的"陵阳公样"，构成式样和组织排列均带有波斯风格。此类纹样从题材到图案变化手法，也深受西亚和拜占庭帝国艺术风格的影响。《唐六典》提到的"蕃客锦袍"、阎立本《步辇图》中的来使袍饰，都可在阿拉尔墓中的袍料中找到。可见，北宋的织锦在内容和风格上仍继承着隋唐的传统风格。

By the analysis of brocade robes of the Northern Song Dynasty unearthed from Alaer City, Xinjiang, the brocades used for the robes include the brocade with patterns of linked pearls and double birds, brocade with patterns of linked pearls, fret and double sheep (Figure 2-9), brocade

图2-9 球路双羊回纹锦（北宋，
新疆维吾尔自治区博物馆藏）
Brocade with Patterns of Linked Pearls,
Fret and Double Sheep（Produced in
Northern Song Dynasty and Collected
in Museum of the Xinjiang Uygur
Autonomous Region）

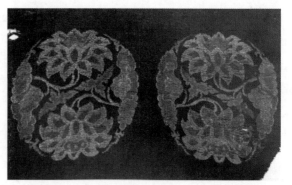

图2-10 重莲团花锦（北宋，故宫博物院藏）
Brocade with Rounded Lotus Patterns（Produced in
Northern Song Dynasty and Collected in the Palace
Museum）

with patterns of vulture and double sheep, brocade with rounded lotus patterns（Figure 2-10）and so on. The patterns of linked pearls, double birds, double sheep and double animals all belong to Lingyanggong patterns popular in the Tang Dynasty, with the Persian style in composition and arrangement. This kind of pattern is also deeply influenced by the artistic style of West Asia and Byzantine Empire from theme to pattern variation. The brocade robes for Fan Ke（foreigners who came to China for commercial trade, diplomatic mission, exchange visit and so on）mentioned in *Six Codes of the Tang Dynasty* and the brocade patterns of the Tibetan envoy's robes in the painting titled *Emperor Taizong Receiving the Tibetan Envoy* by the Tang artist Yan Liben can be verified in the robes unearthed from Alaer. The brocade patterns of Northern Song Dynasty still inherit the traditional style of Sui and Tang Dynasties in theme and style.

（二）宋代后期宋锦的艺术风格/The Artistic Style of Song Brocade in the Late Song Dynasty

宋代中后期，丝绸织锦的艺术风格并没有被唐代装饰性较强的图案风格所束缚，而是在其中注入了富有时代特色的写生风景。当时的画院画风内容追求"诗情画意"，形象刻画细腻、生动，丝绸上不少花鸟形态十分写实。从福建福州黄昇墓、江苏金坛南宋周瑀墓、苏州虎丘塔等地出土的不少丝绸匹料、袍料、残片以及部分宋锦传世品中，均反映出了这种写实的风格，图2-11所示为牡丹芙蓉纹锦。

In the middle and late Song Dynasty, the artistic style of silk brocade was not bound by the decorative pattern style of the Tang Dynasty, but brought with the sketching style with the characteristics of the times. At that time, the Imperial Art Academy advocated that the paintings should be "poetic and picturesque", and the images should be delicate and vivid. Therefore, flowers and birds as brocade patterns were in realistic style. This realistic style is reflected in many silk pieces, robes, fragments unearthed from Huangsheng's Tomb in Fuzhou, Fujian Province,

Zhou Yu's Tomb of the Southern Song Dynasty in Jintan, Jiangsu Province, Huqiu Pagoda in Suzhou, etc., and some products handed down from the Song Dynasty. Figure 2–11 shows the brocade with patterns of peony and hibiscus.

图2-11　牡丹芙蓉纹锦
Brocade with Patterns of Peony and Hibiscus

从历代文献记载，宋代织锦在花纹和色彩上称得上是丰富多彩，名目繁多。董其昌撰写的《筠清轩秘录》中记载了诸多纹样。其他历史文献还有费著的《蜀锦谱》、陶宗仪的《南村辍耕录》、周密的《齐东野语》以及《宋史·舆服志》等，诸书中所记载的名色还有：八答晕锦、云雁锦、真红锦、大窠狮子锦、双窠云雁锦、宜男百子锦、青绿瑞草云鹤锦、真红穿花凤锦、真红雪花球路锦、真红樱桃锦、真红水林檎锦、天马锦、聚八仙锦、宝照锦、灯笼锦、青红捻金锦等。宋代主持茶马贸易的"茶马司"还在四川特设锦坊，织造西北和西南少数民族喜爱的宜男百子锦、大缠枝青红被面锦、宝照锦、球路锦等以交换茶马司所需物资。

According to the historical records, the brocade of Song Dynasty was rich in both pattern and color. *The Secret Records of Jun Qingxuan* written by Dong Qichang depicts abundant pattern designs of Song brocade. In some other historical records, such as *The Book of Shu Brocade* written by Fei Zhu, *Farming in Nan Village* written by Tao Zongyi, *Qidong Yeyu* (the historical recordings of anecdotes) written by Zhou Mi and *The Records of Carriage and Clothing* of *History of the Song*, other famous brocades with various pattern designs were recorded, including the Badayun brocade, brocade with wild goose patterns, red–colored brocade, brocade with large round patterns and lion patterns, brocade with double round patterns and wild goose patterns, brocade with daylily patterns, brocade of turquoise ground with patterns of auspicious grass and crane, brocade of red ground with patterns of phoenix in flowers, brocade of red ground with patterns of snowflake and linked pearls, brocade of red ground with cherry patterns, brocade of red ground with crabapple patterns, brocade with winged horse patterns, brocade with patterns of the Eight Immortals, brocade with patterns of Bao Zhao flowers, lantern brocade, twisted gold brocade, etc. In the Song Dynasty, the Bureau of Tea and Horse, which was in charge of the tea and horse trade, also set up a brocade workshop in Sichuan to weave brocade that was popular among ethnic minorities in northwest and southwest China, such as brocade with patterns of daylily and a hundred boys, brocade quilt of green and red ground with large entangled branches, brocade with Bao Zhao flower patterns and brocade with linked pearls patterns, to exchange for what they needed.

此外，在元宵节时穿着以灯笼为题材的服饰，《邵氏闻见录》记"张贵妃又尝侍上元宴于端门，服所谓灯笼锦者"，织锦、刺绣、缂丝等各种制作手法应有尽有。这类题材一直沿用至明清时期的袍服上，其中有灯笼纹天下乐锦（图2-12）、五谷丰登灯笼纹锦、长寿乐灯

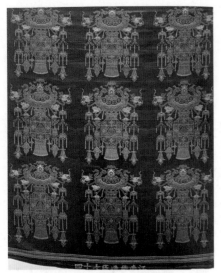

图2-12 灯笼纹天下乐锦
Lantern Brocade with TIAN XIA LE
(a Universal Celebration) Patterns

笼纹锦、双龙灯笼景刺绣圆补、缂丝八团灯笼景锦袍以及江南织造臣七十四灯笼纹锦等。

Women wore clothes with lantern patterns during the Lantern Festival. *Records of Shao* recorded that the imperial concubine Zhang attended the banquet on Lantern Festival and wore the lantern brocade. Various techniques such as brocade weaving, embroidery and kusso (silk tapestry) were used. This kind of theme had been used in the robes of Ming and Qing Dynasties, including lantern brocade with the TIAN XIA LE (a universal celebration) patterns (Figure 2-12), lantern brocade with WU GU FENG DENG (a good harvest) patterns, lantern brocade with CHANG SHOU LE (longevity and blessing) patterns, round Buzi (the badge decorated on official uniforms) with patterns of double dragons and lanterns, kusso robe with patterns of eight rounded flowers and lanterns, lantern brocade with Chinese characters of "JIANG NAN ZHI ZAO CHEN QI SHI SI (Qi Shisi, the official of Jiangnan Weaving Bureau)", etc.

另外，宋金时期，回鹘人擅长织金工艺，并向中原地区传授了这种织造技术，故在宋锦中加金线以及衣服以金为饰的风气在当时大为流行，这样就使部分宋锦显得光彩夺目，富丽堂皇。如前文所提及的故宫博物院收藏的"盘绦花卉锦"就是这种加织金线的宋锦。

In addition, during the Song and Jin Dynasties, the Huihu people were good at gold weaving and taught this skill to people in the Central Plains. Therefore, the trend of adding gold threads to Song brocade and decorating clothes with gold was very popular at that time, which made some Song brocade glittering and resplendent. Song brocade with patterns of flowers and colorful silk ribbons collected in the Place Museum, which is mentioned above, just belongs to this kind.

（三）宋锦的独特之处和名作介绍/The Uniqueness of Song Brocade and an Introduction to Some Famous Works

1. 宋锦的独特性/The Uniqueness of Song Brocade

宋锦较汉锦和唐锦在组织结构和艺术风格上都有很大的突破和创新。

Compared with Han brocade and Tang brocade, Song brocade has great breakthrough and innovation in fabric weave and artistic style.

首先，在织物结构上，改变了汉代经锦仅以经线显花和唐代纬锦仅以纬线显花的局限性，采用了经纬线联合显花的组织结构，使织物表面色彩和组织层次更为细腻和丰富，这是划时代的突破。

Firstly, in the fabric weave, not as the warp brocade displaying patterns through warp in the Han Dynasty and the weft brocade in the Tang Dynasty by weft, it adopted the fabric weave that displayed patterns by both warp and weft, which made the surface color and fabric gradation more delicate and richer. It can be regarded as an epoch-making breakthrough.

其次，在丝线材料上，宋锦采用了一组较为纤细的经线（称接结经或面经），来接结织物正反两面长浮长的纬线，使织物花纹更为清晰、丰满、肥亮，质地又较经锦和纬锦轻薄，更适用于服饰和书画的装裱、装帧，这是厚重的汉锦、唐锦以及云锦所不及的。

Secondly, in silk material, Song brocade adopted a set of slender warp（called stitching warp）to connect the long and floating weft threads on the front and back sides of the fabric, which made the pattern clearer, fuller and brighter; and the texture was lighter than warp brocade and weft brocade, more suitable for mounting and binding clothing, paintings and calligraphy than Han brocade, Tang brocade and Yun brocade.

在制作工艺上，主要应用了彩抛换色的独特工艺，传统称"活色"技艺，即在不增加纬线重数和织物厚度的情况下，使织物表面色彩多变而丰富，甚至可以做到整匹锦的花纹色彩均不相同。此工艺特征不但被云锦所吸收和发扬，而且保留到当代的织锦工艺上，如图2-13和图2-14所示。

图2-13　活色生香的宋锦（正面）　　图2-14　活色生香的宋锦（反面）
Song Brocade with Vivid Colors（Front）　Song Brocade with Vivid Colors（Back）

Thirdly, in the production process, the unique technology of throwing and color changing （traditionally called "HUO SE"）was adopted, that is, without increasing the weft number and fabric thickness, the surface color of the fabric was changeable and rich, and even the pattern colors of the whole brocade could be different. This technology was not only absorbed and developed by Yun brocade, but also borrowed by contemporary brocade weaving, as shown in Figures 2-13 and 2-14.

再次，在图案风格上，它以变化几何图案为骨架，如龟背、四答晕、六答晕、八答晕等，内填自然花卉、吉祥如意纹等，配以和谐的地色，略加对比色彩的主花，使之艳而不俗，古朴高雅。既具有唐宋以来的传统风格特色，又与元明时期流行的光彩夺目的织金锦、妆花缎等品种有着明显的区别，更符合贵族和士大夫阶层崇尚优雅秀美的艺术品位，如图2-15和图2-16所示。

图2-15 绿艾地八答晕纹锦
Brocade of Green Ground with
Badayun Patterns

图2-16 龟背龙纹宋锦
Song Brocade with Tortoiseshell and
Dragon Patterns

In terms of pattern style, it took changing geometric patterns as the skeleton, such as tortoiseshell, Sidayun, Liudayun, Badayun, etc., filled with flowers or auspicious patterns, etc., matched with harmonious ground color and main flowers of slightly contrasting colors, making the brocade simple and elegant. The brocade had the traditional style since the Tang and Song Dynasties, but was obviously different from the dazzling gold brocade and Zhuang Hua Silk popular in the Yuan and Ming Dynasties, which was more consistent with the artistic taste of aristocrats and scholar-officials advocating elegance and grace, as shown in Figures 2-15 and 2-16.

　　最后，在织物用途上，宋锦由于质地较轻薄、精细，风格又古朴典雅，故用途广泛，除适宜制作服饰品以及屏风、靠垫、坐垫等装饰品外，也适用于书画、挂轴、锦匣的装帧，如图2-17～图2-19所示。

Finally, in terms of fabric usage, Song brocade is suitable for making clothing and decorations such as screens and cushions, and is also used for the mounting of paintings, calligraphy, hanging scrolls or brocade boxes, as shown in Figures 2-17 to 2-19.

2. 宋锦纹样名作介绍/An Introduction to Some Famous Works

　　宋锦纹样大多以满地规矩几何纹为特色，其造型繁复多变，构图纤巧秀美，色彩古朴典雅，与唐锦讲究的雍容华贵形成了明显的对比。明清时宋锦的纹样以追慕宋代织锦的艺术格调为特色，但由于宋锦的品种类别不同，其使用功能各有侧重，故纹样形式和题材各有其独特之处。

Most of the patterns of Song brocade are characterized by regular geometric patterns all over

图2-17 宋锦桌旗
The Table Flag Decorated with Song Brocade

图2-18 宋锦抱枕
Song Brocade Used for Cushions

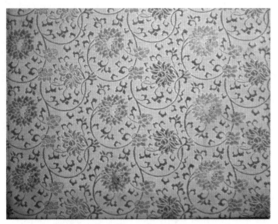

图2-19 宋锦包首
Song Brocade Used for Scroll Mounting

the ground, with complicated and changeable shapes, delicate and beautiful composition, as well as simple and elegant colors, which is in obvious contrast with the rich and magnificent Tang brocade. The patterns of Song brocade in the Ming and Qing Dynasties are featured by following the artistic style of Song Dynasty. However, due to the different varieties and usage, the patterns and themes of Song brocade in the Ming and Qing Dynasties have their own characteristics.

（1）独幅装饰艺术锦。独幅装饰艺术锦主要形式为挂轴、壁毯、卷轴等，专供宫廷陈设用，因而图案多为佛像、经变故事画和花鸟画等。这类宋锦制作精良、纹样写实、气魄宏伟。例如，乾隆时期的重锦《西方极乐世界图轴》(图2-5)，高448cm，宽196.5cm。它以佛祖阿弥陀佛为中心，在佛光放射、祥云缭绕、宫殿巍峨、宝池树石、奇花异鸟的环境中，安排了278位神态各异的人物像，分成上段、中段和下段。上段织的是富丽庄严的殿宇，屋顶上放射出10道佛光，佛光上派生出28尊佛像，象征佛祖在天界化成千佛；中段织的是阿弥陀佛坐于正中，观音和势至两尊菩萨紧侍左右，周围围护着供奉菩萨、天王神将、罗汉众僧、歌伎乐师等；下段织的是九尊莲池，九位转生的人各跪于一朵莲花上，由于生前善恶不同，延伸成九品，上善都转生为佛。画幅边饰及上下裱首和绶带装饰均一气呵成。这幅图轴需用特阔织机，由4~5位工匠同时制织，整幅织物效果极其富丽堂皇。图案风格上、织物结构上以及制作工艺上，均表现了高超的技巧，是科学性与艺术性的完美结合，堪称稀世珍宝。

Single piece decorative and artistic brocade. The main forms of the single piece decorative and artistic brocade included hanging scrolls, tapestries, scrolls, etc., which were specially used for the interior furnishing of the imperial court, so the patterns were mostly Buddhist figures, Buddhist stories, flower-and-bird paintings. This kind of Song brocade was well-made with realistic patterns, imposing and magnificent. *The Sukhavati Brocade Scroll* during the reign of Qing Emperor Qianlong（Figure 2-5）is one of these. It was 448 centimeters high and 196.5 centimeters wide. With Amitabha Buddha as the center, the glorious Buddha's Light, clouds, towering palaces, pools, trees stones, flowers and birds as the background, it arranged 278 Buddhist figures with different facial expressions, which were separately set into the upper, middle and lower parts. In the upper part, ten streaks of Buddha lights were radiated from the roof of magnificent and solemn temple, and from each 28 Buddha statues were derived, symbolizing that Buddha turned into thousands of Buddhas in heaven. In the middle part, Buddha Amitabha sat in the middle, Guanyin（Avolokitesvara）Bodhisattva and Shizhi（Mahasthamaprapta）Bodhisattva stood on the left and right, surrounded by other Bodhisattvas, Lokapalas, Arhats, musicians, etc. In the lower parts, there were nine reincarnated people, each kneeling on a lotus flower in a lotus pond. According to the Buddhist stories, those extremely virtuous could be reincarnated as Buddhas after their death. The margins, upper and lower mounting head and decorative ribbons were woven wholly without stop. *The Sukhavati Brocade Scroll* was woven on an extra-wide loom by 4-5 craftsmen at the same time. It shows magnificent fabric effect, and displays superb skills in pattern style, fabric weave and production technology, which is a perfect combination of science and artistry, and can be called a rare treasure.

该幅巨作封存在故宫博物院已有300余年，从未公开展示。如今由宋锦织造技艺的国家级传承人钱小萍担任技术总设计，与南京云锦研究所共同合作，根据图片资料研究和创制，历经数年，终于在手工花楼机上成功制织，再现了该巨幅宋锦的辉煌。

This masterpiece has been preserved in the Palace Museum for more than 300 years and has never been publicly exhibited. Designed by Qian Xiaoping, the national inheritor of Song brocade craftsmanship, with the cooperation of Nanjing Yun Brocade Research Institute, after several years' efforts, the replica of *The Sukhavati Brocade Scroll* was successfully woven on the manual Hualou loom, which reproduced the glory of this masterpiece.

（2）几何散花匹料锦。这类宋锦是宋锦中实用性较广的品种，可供珍品书画装裱、经卷裱封、幔帐、被面、垫面以及衣料等用，如图2-20和图2-21所示。

Brocade with patterns of geometric schemes and scattered flowers. This kind of Song brocade

图2-20　蓝色地福寿三多龟背锦
Brocade of Blue Ground with Patterns of Fingered
Citron, Peach, Pomegranate and Tortoiseshell

图2-21　彩织曲水地鱼藻纹锦
Multicolor Brocade of Flowing Water Ground
with Fish and Algae Patterns

is a variety with wide functions and practicability in Song brocade, which can be used for mounting precious paintings, calligraphy and scriptures, making curtains, quilt cover, cushions and clothing, as shown in Figures 2–20 and 2–21.

宋锦中应用较广的题材除花卉外，以几何纹样居多，其中最具特色的纹样是用几何网架构成的龟背、四答晕、六答晕、八答晕、天华纹、方棋格子及以圆形交切组成的球路纹和以圆形交叠组成的盘绦纹等。这类几何纹都是以垂直线、水平线和对角线组成的米字格作骨架，在垂直方向和对角线交叉的中心点套以圆形和方形，再在圆形和方形范围内填绘自然形花朵和其他几何纹。骨架线向上下左右及斜角八个方向相连的称为八答晕锦（图2-22）。

Besides flowers, geometric patterns are widely used theme in Song brocade, among which the most distinctive patterns are tortoiseshell, Sidayun, Liudayun, Badayun, Tianhua, square grids, which are all formed by geometric grids, linked pearls composed of intersecting circles and silk ribbons composed of overlapping circles, etc. In this kind of geometric pattern, grids composed of vertical lines, horizontal lines and diagonal lines are used as the skeleton, circles and squares are set at the center point where the vertical lines and diagonal lines cross, and then natural flowers and other geometric patterns are filled in circles and squares. The brocade with the

图2-22　八答晕锦
Badayun Brocade

patterns in which the skeleton lines is connected in eight directions: up, down, left, right and four oblique angles, is called Badayun brocade（Figure 2-22）.

另外，重锦织成的铺垫、靠背和迎手等是根据宫殿中各式龙椅、宝座的实际尺寸设计的，其纹样风格与宫殿内的环境、设施相协调，纹饰题材多为云龙、云蝠、夔龙、球路及缠枝宝相花等。

In addition, the seat cushions, back cushions and arm pillow woven with double brocade were designed according to the actual sizes of various the emperor's chairs and thrones in the palace, and their pattern styles were in harmony with the indoor settings of the palace. The decorative themes were mostly dragon, bat, kuilong（a single legged mythical animal）, linked pearls, entangled Baoxiang branches and flowers, etc.

3. 宋锦的主要价值/The Value of Song Brocade

基于宋锦所具有的上述特点，自宋代起，它便取代了秦汉时期的经锦、隋唐时期的纬锦，在宋元明清时期蓬勃发展，可见宋锦有其杰出的价值，主要表现在以下几个方面。

Based on the above characteristics of Song brocade, since the Song Dynasty, it had replaced warp brocade in the Qin and Han Dynasties and weft brocade in the Sui and Tang Dynasties, and flourished in the Song, Yuan, Ming and Qing Dynasties. Song brocade has its outstanding values as follows.

（1）历史价值。宋锦源于春秋，形成于宋代，辉煌于明清，是中国丝绸传统技艺杰出的代表作之一，也是苏州这座古城特有的人类非物质文化遗产代表作之一。苏州宋锦在历史上一直处于领先地位。宋锦在公元1465～1505年，即明、清两朝为繁盛时期，苏州织造府织造的龙衣、帛、锦、纱、缎、绢等，以宋锦最为著名。苏州当时作为全国丝织业的中心，官办、民办产销两旺，盛极一时，有"东北半城，万户机声"之称。苏州商业的繁荣，政治、经济、文化地位的提高，均与宋锦业的发达分不开，故传承宋锦有着深远的历史意义。

Historical value. Song brocade originated in the Spring and Autumn Period, formed in the Song Dynasty, and flourished in the Ming and Qing Dynasties. It is one of the outstanding representative works of Chinese silk traditional craftsmanship and one of the unique representative works of human intangible cultural heritage in Suzhou. Suzhou Song brocade has always been in a leading position in history. Song brocade flourished in the Ming and Qing Dynasties from 1465 to 1505 AD, and Song brocade was the most famous dragon among the imperial robes, brocade, plain gauze, satin and tough silk woven by Suzhou Weaving Bureau. Suzhou, as the center of silk weaving industry in China at that time, was flourishing in government-run and private production and marketing. The prosperity of commerce and the rise of political, economic and cultural status of Suzhou City, are all inseparable from the development of Song brocade industry, so it is of far-reaching historical significance to inherit Song brocade.

（2）科学与艺术价值。宋锦在织物结构上的突破、工艺技术上的变革、艺术风格上的创新，充分显示了它的优越性和杰出性，并以其独特的结构、精湛的技艺、典雅的图案色彩、古朴高贵的艺术魅力在国内外享有盛誉。至今有很多宋锦的精品传世。尤其是故宫博物院收藏的国家一级文物《西方极乐世界图轴》，它必须由2.5米左右的特阔机，6～8个工匠，采用19把长短抛梭，才能织出近200cm宽和450cm高的巨幅作品，其结构之巧妙、工艺之精湛、生产技艺之高超，在当今都难以达到。另外故宫博物院收藏的明代宋锦"盘绦花卉锦"等，即为采用"活色"彩抛工艺的典范，其科学和艺术价值也是不可估量的。

Scientific and artistic value. Song brocade's breakthrough in fabric weave, technological reform and artistic style innovation, which fully shows its prominence. Because of its unique texture, exquisite craftsmanship, elegant pattern colors, simple and noble artistic charm, it enjoys a high reputation at home and abroad. Up to now, many fine works of Song brocade have been handed down from ancient times. In particular, *the Sukhavati Brocade Scroll* collected by the Palace Museum, among the top national treasures of China, must be woven by an extra-wide loom of about 2.5 meters, with 19 long and short shuttles, and by 6–8 craftsmen, so as to finish the huge work with a width of nearly 200 centimeters and a height of 450 centimeters. Its ingenious fabric weave, exquisite craftsmanship and superb skills are difficult to achieve even today. In addition, the Song brocade of Ming Dynasty collected by the Palace Museum, such as "Song-style brocade with patterns of flowers and silk ribbons", is a model of adopting HUO SE technology (throwing and color changing), with immeasurable scientific and artistic value.

（3）应用价值。宋锦自古以来，历经千年演变，但始终是深受国内外欢迎的传统产品。由于其独特的风格、精美的品质，故有着广泛的用途。据《姑苏志》中说，明代宣德年间，曾织《昼锦堂记》及有词曲文字的欣赏品锦，还有紫白落花流水的装裱用锦。成化、弘治年间，进入繁荣时期，锦袍的穿着范围从内府上用扩大到官用，品种较多。到清初织造局生产的正运缎匹分为内府上用和官用两种。

Application values. Song brocade has evolved for thousands of years since ancient times, but it has always been a popular traditional product at home and abroad. Because of its unique style and exquisite quality, it has a wide range of uses. According to *The Local Chronicles of Gusu*, during the reign of Ming Emperor Xuande, *The Story of Zhou Jin Hall*, and some poems were woven as patterns on brocade; there was also brocade in purple and white color with patterns of drifting petals and flowing water used for mounting. During the reign of Ming Emperor Chenghua and Hongzhi, Song brocade entered a prosperous period, and brocade robes were worn from the imperial palace to official use, with many varieties. By the early Qing Dynasty, the brocade produced by the Weaving Bureau fell into two types—for the use of imperial palace and for official use.

尤其在御用贡品中，从帝王后妃的御用服饰，到宫廷帷幔垫榻的装饰；从内廷贵族文人书画、寺庙佛经的装裱、装帧，到对外群臣使节的馈赠礼品等，处处都用到宋锦。当今相信随着人类物质、文化精神生活的提高，宋锦在高档消费方面，将会有一定的市场需求和应用潜力，值得进一步挖掘和开发。

From the royal costumes to the decoration of the palace curtains and couches, from the mounting of paintings, calligraphy and Buddhist scriptures to the gifts given to foreign ministers and envoys, Song brocade was used everywhere. Today, with the improvement of people's material and cultural lives, Song brocade will have greater potential of market demand and application in high-grade consumption, which is worth better exploitation.

第三节　宋锦的传承/The Inheritance of Song Brocade

一、宋锦的抢救和保护/The Rescue and Protection of Song Brocade

对于濒临失传的宋锦织造技艺，如何进行抢救、保护和传承，是摆在我们面前的一个重要课题。抢救、保护和传承是互为因果关系的一个整体，抢救和保护的目的是传承，传承是最好的抢救和保护。所以首先必须将已失传的传统技艺迅速加以抢救和挖掘，并做到逐一抢救、逐一保护和逐一传承。

How to rescue, protect and inherit the weaving craftsmanship of Song brocade, which is on the verge of being lost, is an important issue before us. Rescue, protection and inheritance are an integration with mutual causal relationship. The purpose of rescue and protection is inheritance, and inheritance is the best way of rescue and protection. Therefore, first, we must rescue and excavate the lost traditional craftsmanship quickly, and rescue, protect and inherit them one by one.

当务之急，必须组织力量，多渠道地收集、挖掘和抢救已散失的宋锦技术档案、技术资料和样品，并通过调研、走访，全面了解宋锦有史以来的花色品种类别、生产经营状况、社会需求和市场情况以及现存宋锦技术人才等。

It is urgent to organize forces to collect, excavate and rescue the lost technical archives, technical data and samples of Song brocade through multiple channels, and fully understand the historical varieties, production and operation status, social needs and market conditions, as well as the existing technical talents of Song brocade through investigation and interviews.

其次，必须将收集到的珍贵的宋锦真品加以科学的复制。在复制过程中，不但能探索其技术奥秘，重现其风采，还可复原宋锦织机，培养和锻炼技术人才，这样才能真正做到对宋锦织造技艺的传承和保护。

Secondly, the precious original samples of Song brocade must be scientifically replicated. In the process of replication, we can not only explore its technical mystery and reproduce its charm, but also restore Song brocade looms and train technical talents, so as to truly inherit and protect Song brocade weaving craftsmanship.

二、宋锦的传承和弘扬/The Inheritance and Promotion of Song Brocade

近年来，党和国家十分重视历史文化遗产和非物质文化遗产的保护。自2004年起，由中华人民共和国文化和旅游部等各级政府部门对非物质文化遗产组织申报、评审和列项，几经努力，2006年"宋锦织造技艺"被列为首批国家非物质文化遗产名录，2009年，又被纳入联合国"人类口头与非物质文化遗产"代表作名录。钱小萍被评为宋锦织造技艺首批国家级传承人，沈惠和朱云秀分别被评为省级和市级传承人。

In recent years, the Party and the state have attached great significance to the protection of historical and cultural heritages and intangible cultural heritages. Since 2004, the Ministry of Culture and Tourism of the People's Republic of China and other government departments at all levels have organized the application, review and approval of intangible cultural heritage. "Song brocade weaving craftsmanship" was included on the National Intangible Cultural Heritage List in 2006, and then in the UNESCO Oral and Intangible Cultural Heritage List of Humanity in 2009. Qian Xiaoping was recognized as the first batch of national inheritors of Song brocade weaving craftsmanship, while Shen Hui and Zhu Yunxiu as provincial and municipal inheritors respectively.

2006年，"苏州钱小萍古丝绸复制研究所"成立，这是在当时苏州丝绸博物馆书记陶苏伟的提议和支持下，钱小萍才下决心，以近七十的高龄继续坚守古丝绸研究和复制这块阵地，并重点致力于宋锦的研究、保护和传承。2008年，中共苏州市委宣传部、苏州市劳动和社会保障局、苏州市文化广电新闻出版局、苏州市经济贸易委员会和苏州总工会联合授予"钱小萍宋锦织造技能大师工作室"的铜牌，如图2-23所示。

图2-23 "钱小萍宋锦织造技能大师工作室"铜牌
Bronze Signboard of "Qian Xiaoping's Master Studio of Song Brocade Weaving Craftsmanship"

In 2006, Suzhou Qian Xiaoping's Institute of Ancient Silk Replication and Research was established. It was with the proposal and support of Tao Suwei, then secretary of Suzhou Silk Museum, that Qian Xiaoping made up her mind to stick to the ancient silk replication and research at the age of nearly 70, and devoted herself to the research, protection and inheritance of Song brocade. In 2008, the Propaganda Department of the CPC Suzhou Municipal Committee of China, Suzhou Labor and Social Security Bureau, Suzhou Bureau of Culture, radio television, press and publication, Suzhou Economic and Trade Commission and Suzhou Federation of Trade Unions jointly awarded the bronze signboard of "Qian Xiaoping's Master Studio of Song Brocade Weaving Craftsmanship", as shown in Figure 2-23.

在2014年11月召开的亚太经济合作组织（APEC）峰会上，领导人穿着由宋锦面料制作的"新中装"亮相，使"宋锦"二字几乎一夜之间传遍世界。那么，如何使宋锦得到更好的传承和弘扬？

At the APEC meeting held in November 2014, the leaders were all wearing "new Chinese-style outfits" made of Song brocade fabric, which made Song brocade known all over the world almost overnight. Thus, how can we inherit and promote it?

宋锦的创新，可以从图案风格、织物结构、织造工艺以及产品用途等方面进行改革和创新，但必须是以传统为基础的创新。多年来以钱小萍为代表的技术人员曾设计了不少宋锦艺术品，如宋锦唐卡、宋锦"枫桥夜泊"挂轴、宋锦"百子嬉春"挂轴、宋锦"杨枝观音"和宋锦"品字牡丹"宫扇等。这些作品有别于传统小型四方连续纹样的实用宋锦，而是多姿多彩的大幅独立纹样的宋锦艺术品，但它的结构、工艺以及精美高雅的基本格调没有变。

Song brocade can be reformed and innovated in pattern style, fabric weave, weaving technology and product use, but it should be based on tradition. Over the years, technicians represented by Qian Xiaoping have designed many Song brocade artworks, such as Song brocade with Tangga pattern, Song brocade with the picture of "a night mooring at the Maple Bridge", Song brocade scroll with the picture of "hundred boys playing in spring", Song brocade scroll with the picture of "Guanyin Bodhisattva with willow branches" and Song brocade palace fan with

图2-24　钱小萍设计的"卍"字纹宋锦
Song Brocade with Swastika Patterns
Designed by Qian Xiaoping

peony patterns, etc. Different from those practical brocade works with small square continuous patterns, these are works with large independent patterns, but its fabric weave, craftsmanship and elegant style have not changed.

在APEC峰会上，领导人穿的"新中装"面料，也可以说是宋锦的一种创新，但它不具代表性。2014年初，万事利集团接到APEC峰会领导人服装设计邀请函，公司总裁李建华立即组织设计团队，并决定融入多种丝绸非物质文化遗产项目。钱小萍因担任了杭州万事利集团的技术顾问，所以应万事利之邀，为APEC峰会领导人的服装设计了宋锦面料。考虑到这次对服装面料的要求是高雅而不张扬，低调而不奢华，所以特意设计了"卍"字纹宋锦（图2-24、图2-25）。

At this APEC meeting, the fabric of "new

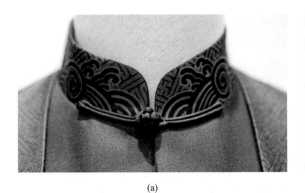

(a)

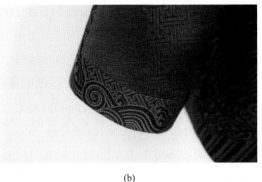

(b)

图2-25　"卍"字纹宋锦新中装
New Chinese-style Outfits of Song Brocade with Swastika Patterns

Chinese-style outfits" worn by leaders can also be regarded as an innovation of Song brocade, but it is not representative. At the beginning of 2014, Wensli Group received the leaders' costume design invitation of Asia-Pacific Economic Cooperation. Li Jianhua, president of Wensli Group, immediately organized the design team and decided to integrate various silk craftsmanship, listed on the Representative List of the Intangible Cultural Heritage of Humanity, into the costume design. As a technical consultant of Hangzhou Wensli Group, Qian Xiaoping designed the fabric of Song brocade for APEC leaders' clothing. Considering that the requirements that the clothing fabrics should be elegant, modest and decent, Qian Xiaoping designed the Song brocade with swastika patterns（Figures 2-24 and 2-25）.

　　新中装的设计团队，在众多面料和式样中博采众长，面料纹样不仅选用了传统宋锦的"卍"字纹，还有"海水江崖"等中国传统图案，并由苏州吴江鼎盛丝绸有限公司负责生产加工。根据北京的天气情况，以真丝线作经，纬线加入羊毛，制织出了这种特殊的宋锦面料。所以，它并不是典型的宋锦，而是在特定要求下的一种创新宋锦。

　　The design team of new Chinese-style outfits learned from many fabrics and styles, selected not only the swastika patterns of traditional Song brocade, but also some other Chinese traditional patterns such as "the sea and mountain pattern". The fabrics were produced and processed by Wujiang Dingsheng Silk Co., Ltd. in Suzhou. According to the weather conditions in Beijing, this special Song brocade fabric was woven with silk thread as warp thread and wool as weft thread. Therefore, it is not a typical Song brocade, but an innovative Song brocade under specific requirements.

　　基于宋锦织物具有质地轻薄细腻，锦面光泽柔和，风格古朴高雅等特点，故较其他织锦具有更广泛的用途。如吴江鼎盛丝绸有限公司将宋锦面料大胆地制作成各式箱包，打破了以往宋锦仅用于服饰、书画和装裱等的局限，这是宋锦在用途上的创新，值得进一步推广。

　　Because Song brocade fabric has the characteristics of light and delicate texture, soft luster, simple and elegant style, etc., it has more extensive uses than other brocade. For example, Wujiang Dingsheng Silk Co., Ltd. boldly made Song brocade fabrics into various bags, which is an innovation in the use of Song brocade and is worth popularizing.

尽管宋锦技艺的传承和弘扬任重而道远，但现在人们已看到濒临衰落和湮没的宋锦终于开始在苏州复苏和新生。尤其是这次APEC峰会领导人"新中装"采用宋锦面料后所产生的影响，相信在各级政府对文化和丝绸的重视下，宋锦文化和技艺将会得到更好的传承和弘扬；有志于从事宋锦事业的人们必将迎来锦样年华，锦样人生。

Although there is a long way to go to inherit and promote Song brocade weaving craftsmanship, people have seen that Song brocade, which is on the verge of decline and oblivion, has finally begun to recover and revive in Suzhou. In particular, the adoption of Song brocade fabrics in new Chinese-style outfits on 2014 APEC. We are convinced that with the attention of governments at all levels to culture and silk, Song brocade culture and craftsmanship will be better inherited and promoted. People dedicated to the cause of Song brocade will surely usher in a golden age of brocade.

◯ 第三章

云锦
Yun Brocade

第一节 **云锦的起源与发展/The Origin and Development of Yun Brocade**

一、云锦的起源/The Origin of Yun Brocade

南京云锦作为中国古代纺织技术的最高成就而享誉中外，其织造工艺已经入选国家级和世界级非物质文化遗产名录。然而，对于云锦的起源，特别是出现年代的界定，中国纺织史学界却众说不一。

Nanjing Yun Brocade, as the highest achievement of ancient Chinese textile technology, is well-known both at home and abroad, and its weaving technology has been included in the lists of national and world-class intangible cultural heritages. However, Chinese textile historians have different opinions on the origin of Yun brocade, especially on the exact year when it was produced.

黄能馥先生（曾先后9次荣获国家级书刊大奖，参与美国耶鲁大学出版社《中国丝绸艺术》编著工作，并在中国丝绸艺术史和中国服装艺术史的研究方面具有突出成就）在2003南京云锦保护国际研讨会上肯定地说："南京云锦的源头发源于公元3世纪的吴国，至公元5世纪刘裕在南京城南的秦淮河畔斗场市（亦名斗场寺）附近设置'斗场锦署'……"

In the 2003 Nanjing Yun Brocade Conservation International Symposium, Mr. Huang Nengfu (who has won the National Book and Periodical Awards for 9 times, participated in the compiling work of *Chinese Silk Art* published by Yale University Press, and has made outstanding achievements in the research of Chinese silk art history and costume art history) affirmed that Nanjing Yun Brocade originated from Wu State in the 3rd century AD, and that Liu Yu, Emperor Wu of Song (Southern Dynasties), set up Brocade Office near Douchang Market (also known as Douchang Temple) on the Qinhuai River in the south of Nanjing in the 5th century AD.

南京大学历史系蒋赞初教授从历史的角度指出：南京云锦的产生和发展既与南京的建都史紧密结合，又与南京作为中国东南地区重镇的历史地位密切相关。蒋赞初教授在《历史事实证明"南京云锦"确属"中华一绝"》一文中，对此做了更为详细的解读："西晋灭吴以后，东吴皇家的织室虽被撤销，但江南民间的丝织业仍然很盛……至西晋覆亡时，中原地区又有近百万人民渡江南下，其中既有上层的皇族和官僚，也有下层的农民和手工业工匠，很可能包括了一部分丝织业工人。而东晋在建康（今南京）建都后不久，就设置了'织坊'，以生产皇室所需的高级丝织品。据《晋书·后妃传》记载，东晋孝武帝之母李太后'本出微贱……时为宫人，在织坊中'。"

Professor Jiang Zanchu, Department of History of Nanjing University, pointed out from a historical view that the emergence and development of Nanjing Yun Brocade is closely related to both the capital construction history of Nanjing and the historical status of Nanjing as a major town in Southeast China. He elaborated on this in his article *Historical Facts Prove that Nanjing Yun Brocade is a Unique Craftsmanship of China*: "After the Western Jin Dynasty destroyed Wu, although the royal weaving room of Eastern Wu was abolished, the folk silk weaving industry in the south of the Yangtze River was still flourishing…By the time of the fall of the Western Jin Dynasty, nearly one million people in the Central Plains had crossed the Yangtze River south, including the royal family and bureaucrats, as well as farmers and handicraftsmen, probably including some silk weaving workers. However, shortly after the capital of Jiankang (now Nanjing) was built in the Eastern Jin Dynasty, weaving workshops were set up to produce high-grade silk for the royal family. According to *Biographies of Empresses and Imperial Concubines* in *the Book of Jin*, the Empress Dowager Li, the mother of Emperor Xiao Wu of the Eastern Jin Dynasty, was from humble origins. She was once a maid in the imperial palace working in the weaving workshop.

对于金陵织锦极具历史意义的时代当属东晋晚期。大将刘裕率军北伐，攻灭建都于长安的后秦国，迁汉魏以来集中于长安的包括织锦工匠在内的中原地区百工至建康（今南京），并于义熙十三年（417年）在建康设置"锦署"，因其地处秦淮河南岸的斗场市，史称"斗场锦署"，这对金陵丝织业的发展起着决定性的影响。所以，南朝人山谦之在《丹阳记》中说："斗场锦署，平关右，迁其百工也。江东历代未有锦，而成都独称妙，故三国魏则市于蜀，而吴亦资西道，至是始有之。"斗场市是东晋南朝时建康都城南郊的重要市场之一，因位于名刹斗场寺（又名道场寺）之前而得名。鉴于后秦是继前秦的一个北方少数民族政权，而前秦苻坚曾经短期统一过北方中原地区，所以这批百工中的织锦工匠既承袭了两汉魏晋的传统，又汲取了少数民族统治者喜爱加金织锦的技艺，加金织锦正是日后南京云锦的主要特征之一。因此，我们有理由认为东晋晚期刘裕从长安迁来的后秦百工，乃是南京云锦业的先驱者。

The historic era for Jinling brocade is the late Eastern Jin Dynasty. Liu Yu, the later Emperor Wu of Song and a general at that time, led the Northern Expedition to attack and destroy the Post-Qin state with its capital in Chang'an, and relocated hundreds of workers, including brocade craftsmen, who came from all over the Central Plains and then concentrated in Chang'an since the Han and Wei Dynasties, to Jiankang. In the 13th year of Yixi (417), a "Brocade Office" was set up in Jiankang, which was called "Douchang Brocade Office" because it was located in Douchang market on the south bank of Qinhuai River, played a decisive role in the development of Jinling silk weaving industry. Therefore, Shan Qianzhi, a county magistrate in Southern Dynasties, once recorded in the document *A chorography of Danyang*: "There was no brocade on the south of Yangtze River in the past dynasties, except for the brocade produced in Chengdu, which can be called a wonder alone. During the Three Kingdoms, both Wei and Wu bought brocade from Shu." Douchang market was one of the important markets in the southern suburbs of Jiankang Capital in the Eastern Jin and Southern Dynasties. It was named after being located in front of the famous Douchang Temple (also known as Daochang Temple). Since the Post-Qin period was a northern minority regime following the Pre-Qin period, and Fu Jian in the Pre-Qin period had unified the northern Central Plains for a short time, the brocade craftsmen during that time not only inherited the traditions of Han, Wei and Jin Dynasties, but also absorbed the skills of gold brocade preferred by ethnic minority rulers. The gold thread added in the brocade was one of the main features of Nanjing Yun Brocade later. Therefore, we have reason to think that the Post-Qin workers who moved from Chang'an by Emperor Wu of Song in the late Eastern Jin Dynasty were the pioneers of Nanjing Yun Brocade industry.

在整个南朝时期（包括宋、齐、梁、陈四朝，420～589年），金陵（当时称建康）的织锦业持续发展。史料记载，南朝各代都在中央政府机构中设立少府，下设平准令以掌织染，锦署亦属平准令管辖，皇室还另设有织室和绣房。南齐时的织成锦工已闻名天下。所以，处于蒙古高原的柔然国曾向南齐政府求取锦工，但被齐武帝以"织成锦工，并女人，不堪远涉"为由，婉言谢绝。梁武帝时，又出现了"公家织官纹锦饰"（《南史·梁本纪上第六》）的规范，侯景曾向梁武帝索锦万匹，为军人袍，虽未获允，也从侧面反映当时御府锦署之产量（《梁书·侯景列传》）。至陈代，地方官员一次进献罗纹锦被达两百条之多（《南史·陈本纪下第十》）。凡此，均可说明南朝织锦业的发达程度。

Throughout the Southern Dynasties (including Song, Qi, Liang and Chen Dynasties, from 420 to 589 AD), the brocade industry in Jinling (then called Jiankang) continued to develop. According to historical records, in each dynasty of the Southern Dynasties, Shaofu (a governmental agency in charge of royal clothes, treasures and food) was set up in the central government agencies, and Pinzhunling (an official post) to handle with weaving, dyeing and brocade. Moreover, the royal family had another weaving room and embroidery room. The brocade weaving workers in the Southern Qi Dynasty had become famous. Therefore, the Kingdom of Rouran, located in the Mongolian Plateau, once asked the government of Southern Qi for brocade workers, but was politely declined by Emperor Wu of Qi on the grounds that "weaving brocade workers include women, so they can't bear the hardship of going far". During the reign of Emperor Wu of Liang, the regulation of "official brocade pattern designs" was proposed (recorded in *The Annals of Liang Dynasty* of *The History of the Southern Dynasties*). Hou Jing, a traitorous general, once asked

Emperor Wu of Liang for a thousand bolts of brocade for military robes (recorded in *A Biography of Hou Jing* of *History of Liang of the Southern Dynasties*). Although his request was not allowed, it also reflects the output of the imperial brocade at that time. By the Chen Dynasty, local officials had tributed as many as 200 brocade quilts at a time (recorded in *The Annals of Chen Dynasty* of *The History of the Southern Dynasties*). All these above can demonstrate the advanced brocade industry in the Southern Dynasties.

从南朝织锦的命名来看，首次出现了"云锦"的名称，见于南朝的文献《殷芸小说》。原文为："天河之东，有织女，天帝之子也，年年织杼役，织成云锦天衣。"作者借神话的叙述，道出了人间巧妇的精湛技艺。结合《齐书·舆服志》"加饰金银薄，世亦谓为天衣"的记载，两处可以互为印证。由此可证，"云锦"之名实始于南朝，而且特指加饰金银箔的织金锦。

As for the naming of the brocade in the Southern Dynasties, the name of "Yun Brocade" appeared for the first time in *Yinyun's Fiction,* a collection of short stories compiled by Liang Yinyun in the Southern Dynasties. According to a fiction in this book, "In the east of the Milky Way, the Weaver Girl, the daughter of the Heavenly Emperor, works hard every year to weave cloud-patterned heavenly brocade clothes." The author narrated the superb weaving skills on earth by myth of fairyland. *The Records of Carriage and Clothing* of *History of Qi of the Southern Dynasties* recorded that "(the robes worn by the emperor were) decorated with thin gold and silver, and called heavenly clothes at that time", which can corroborate each other with the records of *Yinyun's Fiction*. So it can be proved that the name of "Yun Brocade" actually originated in the Southern Dynasties, and it specifically refered to the gold brocade decorated with thin gold and silver.

南唐灭亡后，北宋政府对金陵的政治经济地位相当重视，宋仁宗为皇子时曾被封为升王兼江宁府尹，并以南唐皇宫为府署，著名的政治家包拯与王安石也曾先后担任过江宁府尹。南宋时更定金陵为"行都"，改称建康，以府署为"行宫"，并进一步发展了金陵的丝织业和铸钱业，以适应向金人送"岁币"的需要。建康每年入贡的锦缎等高级丝织品达到上万匹，金陵丝织业的规模至此又有了很大的增长。

After the demise of the Southern Tang Dynasty, the government of the Northern Song Dynasty attached great importance to the political and economic status of Jinling. When Emperor Renzong of the Song Dynasty was the prince, he was made Prince Sheng and prefect of Jiangning, with the Imperial Palace of the Southern Tang Dynasty as the government office. Famous politicians Bao Zheng and Wang Anshi also served as prefect of Jiangning successively. In the Southern Song Dynasty, Jinling was designated as the Temporary Capital and renamed Jiankang, and the government department as the "temporary dwelling place of the emperor", which had promoted silk weaving industry and money casting industry in Jinling, so as to meet the needs of sending "Suibi (the annual tribute of money and goods from the imperial court to foreign nations)" to the Jin Dynasty. The Jiankang government paid tens of thousands bolts of high-grade silk fabrics including brocade as the tribute every year. Therefore, the scale of silk weaving industry in Jinling had increased greatly.

　　两宋以后，南京的丝织业获得了前所未有的恢复与发展。江南丝织业的重心之一——南京，倾城上下，街头巷尾，"百室机房，机杼相和"，以"鸡鸣"为号，昼夜繁忙。仁和里锦绣坊、乌衣巷染坊、靛蓝所、彩帛行、销金铺、吴绣庄等随处可见，水陆埠际，万商云集，呈现一派盛况空前的繁荣景象。图3-1所示为宋代花楼织机，图3-2所示的《耕织图》为南宋时蚕桑丝织生产发展的真实写照。

图3-1　宋代花楼织机
The Hualou Loom in the Song Dynasty

图3-2　《耕织图》
The Farming and Weaving Picture

After the Song Dynasty, the silk weaving industry in Nanjing had got unprecedented recovery and development. Nanjing, one of the centers of the silk weaving industry in the south of the Yangtze River, was busy day and night with "about 100 weaving workshops and numerous looms operating". Brocade workshops, dyeing houses, indigo institutes, color silk stores, gold-embedding shops and embroidery shops were scattered everywhere in the city, and thousands of businessmen gathered, showing an unprecedentedly prosperous scene. The Hualou loom in the Song Dynasty is shown in the Figure 3-1. Figure 3-2 shows *The Farming and Weaving Picture*, a portrait of serioulture and silk weaving production in the Southern Song Dynasty.

　　综上所述，南京云锦的起源上溯到1600年前，即东晋政府在秦淮河畔以长安迁来的百工为主力创建"锦署"（东晋义熙十三年，公元417年）；而织金锦（金银箔）在金陵的开始织造与"云锦"一名在南朝记载中的出现，至少也有1500年的历史。已有的资料证明，南京云锦在元、明、清三代很兴盛。直到现在，其生产流程和技术传统仍没有中断。

　　To sum up, the origin of Nanjing Yun Brocade can be traced back to more than 1,600 years ago, that is, the thirteenth year of Yixi in the Eastern Jin Dynasty (AD 417), when the Eastern Jin government established the brocade office on the Qinhuai River with workers moving from Chang'an as the main force. The weaving of gold brocade (thin gold and silver) in Jinling and the name of "Yun brocade" in the records of the Southern Dynasties have a history of at least 1,500 years. The existing data prove that Nanjing Yun Brocade flourished in the Yuan, Ming and Qing Dynasties. Its production process and technical tradition have continued until now.

二、云锦的繁盛/The Prosperous Period of Yun Brocade

1. 元代为云锦发展奠定了基础/The Yuan Dynasty Laid a Foundation for the Development of Yun Brocade

南京云锦在元代的生产主要是在官办织造机构的主持、管理下进行的。元代在南京设立的官办织造机构名叫东织染局、西织染局。据元至正年间的《金陵新志·历代官制》记载：“东织染局至元（元世祖）十七年（1280年），于城东南隅前宋贡院立局。有印，设局使二员，局副一员。管人匠三千六户，机一百五十四张，额造缎匹四千五百二十七段，荒丝一万一千五百二斤八两。隶资政院管领。”图3-3所示为元代出土的织锦残片。

图3-3　元代出土织锦残片
The Unearthed Brocade Fragments of the Yuan Dynasty

The officially-run weaving institutions mainly conducted and managed the production of Nanjing Yun Brocade in the Yuan Dynasty. The officially-run weaving institutions in Nanjing in the Yuan Dynasty were called East and West Weaving and Dyeing Bureau. According to *The Official System in Past Dynasties* of *New Chronicle of Jinling*, in the 17th year of the reign of the First Yuan Emperor Kublai Khan（1280）, the East Weaving and Dyeing Bureau established a bureau in front of the former examination hall of the Song Dynasty in the southeast corner of the city, with two bureau envoys and one deputy bureau member, in charge of 3,600 households of craftsmen and 154 looms, with annual manufacturing task of about 4,527 bolts of satin and about 11,500 jin（a unit of weight, =1/2 kilogram）raw silk. Figure 3–3 shows the unearthed brocade fragments of the Yuan Dynasty.

“西织染局至元十七年于侍卫马军司立局，设官与东织染局同”。从《元史·百官志》《元史·后妃列传》的记载还可以知道，南京（宋时称建康府、元时改称集庆路）的东、西织染局隶属资政院，于元末时属于顺帝皇后完者忽都位下的，其所织币帛锦缎等丝织品均为中宫所用。这些官办织造机构，每年耗费巨大的人力、物力和财力，生产御用的龙衣、蟒袍和大量的各色花素缎匹，专供皇室、贵族享受，并用以“章贵贱，别等威”和祭祀颁赏之需。

Still recorded in *The New Chronicle of Jinling*, in the same year（1280）, the West Weaving and Dyeing Bureau was set up, and the official posts were the same as the East Weaving and Dyeing Bureau. According to *Official Records* and *The Biography of Empress Concubine* in *History of the Yuan Dynasty*, the East and West Weaving and Dyeing Bureau of Nanjing (Jiankang Prefecture in the Song Dynasty and Jiqing Road in the Yuan Dynasty) was subordinate to the Central Advisory Council, and at the end of Yuan Dynasty, it belonged to Wan Zhehudu (1315–1369), the queen of the Emperor Shundi of the Yuan Dynasty, and woven silk fabrics such as silk and brocade for the harem. These officially-run weaving institutions spent huge manpower, material and financial resources every year to produce imperial robes, official robes and a large number of colored and plain satin, exclusively for the royal family and aristocracies to show their distinguished status and satisfy the needs of sacrifice and reward with clothing.

元代统治者喜爱使用金锦，有爱用金装饰丝、毛织物的习尚，从明、清两代继续作为御用贡品的云锦上，也能明显地看到元代这种用金风气的直接影响。云锦中的"库金""织金锦、缎"等，就是从元代金锦延续发展下来的品种（现在只有云锦中保存着这种织金品种）。用金装饰丝织物纹样的做法，在元代以来的南京锦缎中，得到了继承和发展，成为南京云锦的一个重要装饰特征。此外，生产近一个世纪的元代锦缎，创造了很多优美的串枝花图案和各种云纹图案、吉祥图案，给明、清两代云锦图案的设计留下了丰富的样式和深远的艺术影响。直到现在，串枝花图案（如串枝牡丹、串枝莲）和云纹图案，仍是云锦图案中常用的格式和纹样题材，如图3-4和图3-5所示。

The rulers of the Yuan Dynasty preferred gold brocade, and they loved to decorate silk and wool fabrics with gold. We can clearly see the direct influence of this trend from the Yun brocade which continued to be used as imperial tribute in the Ming and Qing Dynasties. Palace gold brocade (one of the traditional varieties of Yun brocade, with full display of gold as its characteristics), gold brocade and gold satin are all Yun brocade varieties that have continued to develop from gold

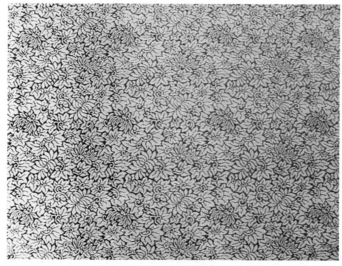

图3-4 织金锦
Gold Brocade

图 3-5　金代齐国王金字袍
The Gold Robe of King Qi in the Jin Dynasty

brocade of the Yuan Dynasty（now these varieties can only be seen in Yun brocade）. The practice of decorating silk fabric patterns with gold has been inherited and developed in Nanjing Yun Brocade since the Yuan Dynasty, and has become an important decorative feature of Nanjing Yun Brocade. In addition, the brocade of Yuan Dynasty, which has been produced for nearly a century, has created various beautiful entangled floral branches patterns, cloud patterns and other auspicious patterns, which have left rich styles and far-reaching artistic influence on the design of Yun brocade patterns in the Ming and Qing Dynasties. Up to now, the entangled floral branches patterns（such as entangled branches of peony or lotus）and cloud patterns are the commonly used formats and patterns in Yun brocade, as shown in Figures 3-4 and 3-5.

2. 明代为云锦的发展构筑了主体/Yun Brocade Gradually Prospered in the Ming Dynasty

　　明代的官营织造，经营单位之多，分布地区之广，规模之庞大，是前所罕见的。按经营体制分，有属中央系统的，有属地方系统的。两京织染和设在南京的"神帛堂"与"供应机房"，都属中央系统管辖。图3-6所示为出土的明代云锦龙袍料。

The official silk weaving industry in the Ming Dynasty had so many operating institutions, so wide distribution and so large scale that were all unprecedented before. In terms of operating system, some institutions belonged to the central system, and others the local system. The Weaving and Dyeing Bureaus in Beijing and Nanjing, the Shenbo Tang（handicraft workshop in the Ming Dynasty）and the Supply Weaving Workshop located in Nanjing were all under the jurisdiction of the central system. Figure 3-6 shows the unearthed apparel fabric of Yun brocade for dragon robe in the Ming Dynasty.

　　所谓"两京织染"，是指分设在南京和北京的两处织染局。设在南京的叫"内织染局"，也叫"南局"，隶属工部。额设织机300余张，有军民人匠3000余名。

图3-6 出土的明代云锦龙袍料
The Unearthed Apparel Fabric of Yun Brocade for Dragon Robe in the Ming Dynasty

Two Weaving and Dyeing Bureaus were set up in Nanjing and Beijing. The one in Nanjing was called "Weaving and Dyeing Bureau for Emperors and Courts", also called "the South Bureau", subordinated to the Ministry of Industry, with more than 300 looms and more than 3,000 military and civilian craftsmen.

设在南京的"神帛堂"属司礼监管辖。据记载，神帛堂额设织机40张，有食粮人匠1200名。每十年织造一次，共织帛料13690段，即平均每年织造神帛1369段。

Shenbo Tang set up in Nanjing was under the jurisdiction of Si Li Jian（an official institution managing eunuchs and palace affairs in the Ming Dynasty）. According to historical records, there were 40 looms and 1,200 craftsmen in it; the silk fabric was woven every ten years, about 13,690 lengths of silk at once, that is, the average annual production was about 1,369 lengths of silk.

"神帛堂"设置的具体地点，据《续纂江宁府志》记载："神帛诰命堂，向在皇城厚载门内。"《同治上江两县志》亦载："往时又有神帛堂，在驻防城北安门内。"据考，厚载门即宫城北门，而北安门才是皇城之北门。以此推断，神帛堂所在的具体地点当在厚载门与北安门之间偏西地区内。

According to *The Successive Compilation of Jiangning Mansion Records*, the specific location of the Shenbo Tang was roughly located in the Houzai Gate of the Imperial City. *Records of Shangyuan and Jiangning Counties During the Reign of Qing Emperor Tongzhi* also contains: "In the past, there was a Shenbo Tang located within the Bei'an Gate of the garrison town." According to textual research, Houzai Gate was the north gate of the Forbidden City, while the Bei'an Gate was the north gate of Imperial City, so it can be inferred that the specific location of Shenbo Tang should be in the western area between the Houzai Gate and the Bei'an Gate.

南京"供应机房"的具体设置地点，据明万历《上元县志》上所附的京城图，上面标示"供应机房"所在方位，即明"汉王"府遗址，约在今逸仙桥和汉府街道迤北和竺桥一带。

According to the capital map attached to *Shangyuan County Records* in the Ming Dynasty, the location of the Supply Weaving Workshop was marked on the site of Prince Han's Mansion, located around the Yixian bridge, Yibei and Zhuqiao in Hanfu street, Nanjing today.

明代的官营织造，每年虽然各有额定的缎匹造解任务，但是由于统治者生活的奢靡和赏赐的无度，每年常额造解的缎匹实际上满足不了这种庞大的消费需要。因此在常额的造解外，往往以"添派"的名目大量增造各种缎匹，其增织的数量，远比正式额定的造解数目大。图3-7 所示为复制的明绛红四合云纹地十团龙袍。

图3-7　明绛红四合云纹地十团龙袍（复制品）
The Dragon Robe (Replica) of Crimson Ground with Moiré Patterns and Ten Dragon Roundel Patterns

Although the official weaving in the Ming Dynasty had an annual manufacturing task of a certain amount of silk, due to the extravagance of rulers' life and excessive rewards, the silk produced could not actually meet this huge consumption demand. Therefore, in addition to the stipulated silk production amount, a large number of silk bolts were often manufactured under the pretext of "extra assignment", which led to an increase in the amount of silks woven, far larger than that of the annual task.

As shown in Figure 3-7, it was a replicated dragon robe of the Ming Dynasty, of crimson ground with moiré patterns and ten dragon roundel patterns.

据有关材料记载，"明代累朝制造缎匹，不过三万匹，上用赏赐，俱在其中"，甚至"一岁所造，供费有余"。

According to official records, during the Ming Dynasty, only 30,000 bolts of satin were produced in a year, among which the silk for the imperial courts and the rewards had been included; the silk and satin annually produced were even more than enough to satisfy all these needs.

然自明朝天顺（1457～1464年）以后，靡费情况日增。如《明史》卷八十二《食货》六记："明初设南北织染局，南京供应机房。各省直岁造供用，苏杭织造，间行间止。自万历中，频岁派造，岁至十五万匹，相沿日久，遂以为常……"

However, since the reign of Emperor Tianshun（1457-1464）, hedonism and extravagance gradually increased. As recorded in *Food and Goods* in Volume 82 of *History of the Ming Dynasty*, in the early Ming Dynasty, the North and South Weaving and Dyeing Bureau and Nanjing Supply Workshop were set up. Each province had an annual manufacturing task to satisfy the consumption of the imperial court, and the weaving in Suzhou and Hangzhou stopped from time to time. Since the reign of Emperor Wanli, the extra task was frequently assigned, and the actual annual manufacturing amount reached 150,000 bolts, which lasted for a long time and finally became a common practice.

明代朝廷根据文武百官职阶的高低，赐给锦缎、纻丝等官服数量，每年就需数万匹之多的锦缎用料。更主要的是，宫廷生活服用的奢靡和最高统治者赏赐的无度，使锦缎耗费的数量极为惊人。这种庞大而无厌的消费需求，往往使官办织造难以胜任和满足。在此形势下，民间的锦缎织造业正是适应这种需要而日渐发展起来的。宫廷和官府除需索于官办织造外，往往采用"领织""收购""采办"等方式，向民间搜罗缎匹，把对锦缎需求的相当一部分转嫁于民间机户的织造上，以弥补官办织造供应之不足。这在一定程度上刺激了民间锦缎织造业的繁荣和发展。

According to the rank of all the officials, the Ming Dynasty court rewarded them a number of official robes made of silk, brocade and satin, which required tens of thousands of bolts every year. What's more, the extravagance of court life and the emperor's reward also consumed a soaring amount of silk and satin. This huge and insatiable consumption requirement often made it difficult for officially-run weaving to satisfy. Under the circumstances, the folk brocade weaving industry developed and prospered just to meet this need. Besides claiming silk and satin from the officially-run weaving institutions, the imperial court and the government often employed the ways of organizing folk workshops to produce, purchasing, procurement and so on to collect silk and satin from the folk, thus transferred a considerable part of the demand to the folk weaving workshops to make up for the shortage of officially-run weaving supply, which had stimulated the prosperity and development of folk weaving industry to a certain extent.

明代宫廷御用的特种锦缎，具有突出成就的是妆花织物。妆花织物的特色是"挖花妆彩"；织造时配色自由，花纹色彩变化丰富。在花纹装饰上，可以逐花异色取得多样而统一的美好效果。这种挖花妆彩的配色技法，在明代以前丝织物花纹的配色上，略作局部的少量运用，可能曾出现过；但作为整件锦缎织料，全部花纹的妆彩方法和整件织料的织造方法，在明代以前的锦缎织物中是不曾有过的。妆花织物是明代早期创造的品种，最初是在缎地提花织物上挖花妆彩，以后把这种配色织造技法，发展运用到纱、罗、绸、绢、绒等不同质地、不同组织的织物上去，达到了无施不巧、非常纯熟的地步，大大丰富了妆花织物的品种内容，把我国彩织锦缎的配色技巧和织造技术发展到一个新的水平。直到现在，这种逐花异色的彩织技术，现代化的电力织机还未能替代。妆花织物是明代南京丝织具有代表水平的产品，它是南京丝织艺人的重大创造和重大贡献。图3-8所示为云锦牡丹妆花缎。

Among the special brocade for the use of the imperial court in the Ming Dynasty, the Zhuanghua silk fabrics were outstanding. The Zhuanghua silk fabrics were characterized by the warp-through and weft-warp swivel weaving technique as well as the free color matching and the changeable pattern colors. On Zhuanghua silk fabrics, the colors of adjacent patterns were different, with the brilliant and harmonious color matching. This color matching of the swivel weaving technique might have already shown up in that of the silk fabric patterns before the Ming Dynasty, as decorations in a very small area; however, it had never been used on a whole brocade

图 3-8　云锦牡丹妆花缎
Zhuanghua Satin with Peony Patterns

fabric before the Ming Dynasty. Zhuanghua silk fabrics were created in the early Ming Dynasty. At first, the swivel weaving technique were only applied on satin jacquard fabrics. Later, it was greatly developed and applied to fabrics with different textures and weaves, such as yarn, half-cross leno, silk, satin, velvet, etc., which reached very fine and highly skillful craftsmanship, greatly enriched the variety and content of Zhuanghua fabrics, and promoted the color matching skills and weaving technology of colored brocade in China to a new level. Until now, modern electric looms have not been able to replace this color weaving technology of different colors. Zhuanghua silk fabric, as a representative product of Nanjing silk weaving of the Ming Dynasty, is a great creation and contribution of Nanjing silk weaving craftsmen. As shown in Figure 3-8, it was Zhuanghua satin with peony patterns.

3. 清代将云锦发展推向了顶端/The Development of Yun Brocade Reached Its Peak in the Qing Dynasty

同明代一样，清代云锦业的发展也是与官营织造分不开的。清代的江宁织造，通常分为两个部分：织造署和织局。"织造署"是督理织造官吏驻扎及管理织造行政事务的官署；"织局"是织造生产的官局作场。前者在今大行宫处；后者又分为"供应机房"和"倭缎机房"。"供应机房"在今汉府街处，"倭缎机房"在今常府街细柳巷口处。

The development of Yun brocade industry in the Qing Dynasty, like that in the Ming Dynasty, was also inseparable from officially-run weaving. Jiangning Imperial Weaving Bureau in the Qing Dynasty consisted of two institutions, the Weaving Department and the Weaving Bureau. The former was the official department to supervise the stationing of weaving officials and manage the weaving administrative affairs, located in today's Da Xinggong（located in the center of the main city of Nanjing）; the latter was the official bureau of weaving manufacturing, which was divided into the Supply Workshop located in today's Hanfu Street, and the Japanese-style Satin Workshop located in today's Xiliu Lane of Changfu Street.

据《大清会典》记载："织造在京有内织染局，在外江宁、苏州、杭州有织造局，岁织内用缎匹，并制帛诰敕等件，各有定式。凡上用缎匹，内织染局及江宁局织造；赏赐缎匹，苏杭织造。"光绪《大清会典》卷1190《内务府库藏》记载："顺治初年（1643年）定，御用礼服，及四时衣服，各宫及皇子公主朝服衣服，均依礼部定式，移交江宁、苏州、杭州三处织造恭进。"由此可见，清代的江宁织局，是织造御用锦缎的主要生产部门。

According to *Compilation of Regulations in the Qing Dynasty*, there were imperial weaving and dyeing bureau in Beijing and weaving bureaus in Jiangning, Suzhou and Hangzhou. The Imperial Weaving and Dyeing Bureau and Jiangning Weaving Bureau were responsible for the manufacturing task of satins used for the imperial court; while the Suzhou and Hangzhou Weaving Bureaus for that of silk for awards to the officials, with the specified formats. As recorded in *The Stock of the Imperial Household Department* in Volume 1,190 of *Compilation of Regulations in the Qing Dynasty*, in the first year of the reign of Qing Emperor Shunzhi（1643）, the royal gowns for the emperor, royal clothes for four seasons, the court dresses for the imperial harems, princes and princesses were all manufactured by Jiangning, Suzhou and Hangzhou weaving bureaus according to the formats stipulated by the Ministry of Rites. Therefore, Jiangning Weaving Bureau in the Qing Dynasty was the main manufacturing department of royal brocade.

江宁织局自顺治二年（1645年）开织，采用的是"买丝招匠"的办法，改变了明末织造"散处民居"的领织经营方式。这样做的原因是鉴于明末的官营织造"无总织局以汇集群工，此明季之所以坐废也"，因而采取集中于织局生产，便于管理的做法。顺治八年（1651年）特下令"织造局照额设钱粮，买丝招匠，按式织造"。尔后，"买丝招匠"便成为清代江南织局经营的定制。

Jiangning Weaving Bureau started weaving in the second year of the reign of Emepror Shunzhi（1645）, and adopted the method of "buying silk and recruiting craftsmen", which changed the leading weaving system in the late Ming Dynasty under which the weaving craftsmen were scattered all over the country. The reason for this was that in the late Ming Dynasty, the officially-run weaving industry had no general weaving bureau to gather craftsmen, so it finally declined. It was more convenient to manage by concentrating the craftsmen under the charge of the general weaving bureau. In the eighth year of the reign of Qing Emperor Shunzhi（1651）, it was specially ordered that "the Weaving Bureau shall buy silk and recruit craftsmen according to the specified amount of money and grain, and weave according to the fixed styles". After that, "buying silk and recruiting craftsmen" became the management system of Jiangnan Weaving Bureau in the Qing Dynasty.

清初的江南织局，未能长年维持生产，常因动乱及兵饷告匮而奉旨停止织造，或奉旨裁减。直至康熙二十五年（1686年）以后，江南各织造的生产方始逐步走上正轨。整个清代以江宁织局的生产为最盛，如图3-9和图3-10所示。

In the early Qing Dynasty, the Jiangnan Weaving Bureau often stopped weaving or cut down by order of the emperor due to turmoil and insufficient soldiers' wages. Until the 25th year of the reign of Emperor Kangxi（1686）, the weaving production of the Jiangnan Weaving Bureau gradually got on the right track. Throughout the Qing Dynasty, the production of the Jiangning Weaving Bureau was the most prosperous, as shown in Figure 3-9 and Figure 3-10.

图3-9　江宁织局匹头（七十四）
The Piece Goods Produced by Jiangning
Weaving Bureau（Seventy-four as the Inspector）

图3-10　江宁织局匹头（贵存）
The Piece Goods Produced by Jiangning Weaving
Bureau（Gui Cun as the Inspector）

织局的织机全部是织局置备的，由织局选择熟悉各项织造业务的匠工领织。匠工来源基本上有以下两种形式：一是通过官府招募，这是织局各色工匠最主要的来源国；二是招收幼匠学艺，成为织局的养成工。除此以外，织局还采用"领机给帖"和"承值应差"的方式，占用和剥削民间丝织手工业工匠的劳动。"领机给贴"是织局发给执照，民间殷实机户领织官局织机。所谓"承值应差"，是官局织造向民间丝织手工业强行摊派的一种无偿劳役。据云锦业的老工人回忆，过去牵经接头在织局应差，只有饭吃，没有钱拿，完全是一种白当差。从实际情况看，"承值应差"的均是民间织造行业中靠出卖手艺劳动为生的技术工匠。根据有关资料对云锦业情况的记载："按锦业最盛时代，当推前清康（熙）、乾（隆）两帝在位时为极。除皇帝、亲王必用外，兼答谢越南、朝鲜等国赠礼之需，复售与住坐南京之富商大贾，运往蒙古及西藏等处。彼时织品尚妆花……此类织法，工精料美，巨细认真，花样繁多，鲜妍夺目。工人不惮思虑、不惜时间，精益求精，日新月异，冀获皇家赏视，博取荣名，是以人争趋之，锦织因以大盛。"在一定程度上也促使了民间丝织手工业的发展。图3-11所示为当时民间机户所织的云锦产品。

The weaving bureau equipped all the looms and selected craftsmen familiar with various weaving businesses. Craftsmen basically came from these two channels. One was through official recruitment, which was the most important source of craftsmen of all kinds in the Weaving Bureau. The other was to recruit apprentices, usually craftsmen's sons, to learn brocade weaving techniques and train them into craftsmen for the Weaving Bureau. In addition, the Weaving Bureau also implemented the ways of "assigning looms to craftsmen and issued licenses to them" and "forcibly apportioning tasks without pay", to occupy and exploit the labor of folk silk craftsmen. According to the old craftsmen's recollection, in the past, they weaved Yun brocade in the weaving bureau with only food but no salary. According to the historical records of the Yun brocade industry, Yun brocade industry witnessed its most prosperous times during the reign of the two emperors Kangxi and Qianlong in the Qing Dynasty; in addition to meeting the daily needs of emperors and princes, Yun brocade can also be used to give Vietnam, Korea and other countries as gifts, or be sold to wealthy merchants in Nanjing, and then

图3-11 民间机户匹头（张象发）
The Piece Goods Produced by Folk Weaving Workshop（Zhang Xiangfa Weaving Workshop）

transported to Mongolia and Tibet. At that time, Zhuanghua brocade was the most fashionable silk fabric, with the meticulous workmanship, exquisite quality, numerous patterns and eye-catching colors. Craftsmen worked hard to keep improving and made progress with each passing day, hoping to gain royal appreciation and win fame and fortune, which led to the flouring of brocade weaving, and promoted the development of folk silk handicraft industry partly. Figure 3-11 shows a Yun brocade product woven by folk weaving workshop at that time.

元、明、清三代南京云锦的历史，就是以官营织造为主线的历史。但是，光绪三十年（1904年）五月二十七日，清光绪皇帝谕："……现在物力艰难，自应力除冗滥，用资整顿……，江宁、苏州两织造同在一省，著将江宁织造裁撤，……以节虚糜，而昭核实。"经历了元、明、清三代、延续达630多年时间的江宁官办织务，从此正式宣告结束。

The history of Nanjing Yun Brocade in the Yuan, Ming and Qing Dynasties was roughly a history of officially-run weaving industry. However, on May 27 in 1904, Emperor Guangxu issued a decree: "Now that material resources were limited, we must vigorously reduce excess and rectify the capital...Since Jiangning and Suzhou weaving bureaus were in the same province, so Jiangning Weaving Bureau will be abolished...to avoid the unnecessary waste." Since then, Jiangning's officially-run weaving business, which had gone through the three historical eras of Yuan, Ming and Qing Dynasties, and lasted for more than 630 years, came to an end.

三、云锦的衰落/The Decline of Yun Brocade

江宁织造局的裁撤，标志着清代在江宁官办手工业的衰落，也标志着云锦在走向衰落。

The abolition of Jiangning Weaving Bureau marked the decline of officially-run handicraft industry in Jiangning in the Qing Dynasty, as well as the decline of Yun brocade.

1911年发生了辛亥革命，推翻了中国历史上最后一个封建王朝，建立了中华民国。从此，兴盛于明、清两代的南京云锦，进入了衰落的历史时期。

The Revolution of 1911 overthrew the last feudal dynasty in Chinese history and established the Republic of China. Since then, Nanjing Yun Brocade, which flourished in the Ming and Qing Dynasties, had entered a historical period of decline.

民国初年，能继承销售关系的，只有内蒙古及西藏部分地区，但只靠一些行商贩运。由于交通条件落后，贩运一次行程往返需8个月的时间，因而每年运销数量甚为有限。据记载，此时全业机户尚存108家，生产资金总额为24519元，织机449台。

In the early years of the Republic of China, brocade was only sold in only parts of Inner Mongolia and Tibet, merely through some itinerant traders. Due to the poor transportation, it took an eight-month round trip to transport goods, resulting in very limited annual sales. According to historical records, there were only 108 loom households in the whole industry at that time, with a total production capital of 24,519 yuan and 449 looms.

1921~1931年，南京云锦行业中曾一度创造了一些结合时代生活的实用新产品。专门向来华旅游的外国人和归国观光侨胞推销，曾经风靡一时，很受欢迎。从创新的意义上来说，这些正是云锦这个传统工艺品为适应现实生活需要，积极开创外销新局面的一个可取的尝试，是探索古为今用的一个良好途径。然而，由于当时的政府对民族工艺的改革和发展采取漠视的态度，以及行业中缺乏专门力量在这方面作悉心的研究和认真的倡导，再加以一些机户的单纯牟利观点，忽视产品质量，以致用料不精、粗制滥造，严重地影响了这些新产品的声誉，终至销路断绝，遭到夭折的命运。

From 1921 to 1931, Nanjing Yun Brocade industry once created some practical new products adapted to the life of the times. It was once all the rage to sell them to foreigners who came to China for a tour and overseas Chinese who returned their homeland for sightseeing. In the sense of innovation, these were all desirable attempts to meet the needs of life and actively create a new situation in the export of traditional handicrafts, and good ways to explore how to make the past serve the present. However, during that time, the Government of the Republic China ignored the reform and growth of national craftmanship; special research and serious advocacy in the industry were inadequate; only searching for profits but ignoring the quality, some loom owners used shoddy materials and manufactured in a rough way. All the factors above seriously affected the reputation of these new products and eventually cut off their sales.

随着20世纪初清王朝的覆灭，手工织造的南京云锦业失去了最大的市场和主要的消费对象，织造者群体与消费群体失去了平衡，云锦行业落入低谷，工匠们纷纷改行。南京很多优秀的传统提花丝织产品，或因生产工艺复杂、工料成本太高难以发展而停产；或因脱离时代需要、脱离人民生活实际而被自然淘汰；也有不少因为人亡艺绝而湮没失传。由于人们对手工承传工艺易丧失的状况没有引起重视，历史上曾出现过的部分品种已失传殆尽。如明代《天水冰山录》（抄宰相严嵩家衣物清单）中记载的妆花绢、妆花绫、妆花绸等名贵品种，现在已只知其名而不知其物了。

With the fall of the Qing Dynasty in the early 20th century, the hand-made Nanjing Yun Brocade industry lost the largest market and the main consumers, and its manufacturing and consumption were out of balance. The brocade industry felled into a trough, and craftsmen had to change careers. The production of many excellent traditional jacquard silk products in Nanjing had to be stopped because of complex production process and high cost of labor and materials; or because it was divorced from the needs of the times and people's lives, it was naturally eliminated. Moreover, some traditional craftsmanship got lost because old craftsmen passed away. Since people had not paid enough attention to the loss of manual craftsmanship heritage, some varieties that once appeared in history fail to be handed down from past generations. For example, in *Tianshui Iceberg Records* (a book written by Wu Yunjia, a scholar in the Qing Dynasty, making a detailed record of the family property of Yan Song, a corrupt official in the Ming Dynasty) are recorded some precious brocade varieties, such as the Zhuanghua lustre, Zhuanghua damask silk and Zhuanghua bourette silk, but up until now people only know their names but don't know what they exactly look like.

1927年国民政府定都南京，据国民政府工商部技术厅的调查，1930年时南京云锦业的机户总数仅47家，机台数为135张，工人总数仅267人。

In 1927, the National Government of the Republic of China set Nanjing as its capital. According to the investigation of the Technical Department of the then Ministry of Industry and Commerce, in 1930, in Nanjing Yun Brocade industry there were only 47 loom households, with 135 looms and 267 workers.

1935年，蒙古人民共和国的"德华洋行"到张家口订购云锦，此时能够承接订货合同的，只有"正源兴"和"恒昌永"两个客帮号家。其余号家，在社会经济不景气的冲击下，均已先后倒闭。曾经发达兴盛的南京云锦业，至此已是一片萧条冷落景象。图3-12所示为正源兴本机妆缎。

In 1935, Deutsch-Asiatische Bank of Mongolian People's Republic went to Zhangjiakou (in Hebei Province today) to order Yun brocade. At this time, only two workshops, "ZHENG YUAN XING" and "HENG CHANG YONG", were able to receive the order. Other workshops, under the impact of social and economic recession, had closed down one after another. Nanjing Yun Brocade industry, which was once prosperous, sank into depression. The satin product manufactured by ZHENG YUAN XING was shown in Figure 3-12.

1941年太平洋战争爆发。在日军的封锁下，海陆交通中断，销路全部断绝，号家关闭，

图3-12 正源兴本机妆缎
The Satin Product Manufactured by
ZHENG YUAN XING

工人失业，云锦的生机全部断绝。

After the Pacific War broke out in 1941, under the blockade of the Japanese aggressors, sea–land traffic was interrupted, all sales were cut off, all brocade workshops were closed, the workers lost their jobs, and the vitality of Yun brocade industry was completely cut off.

1945年抗日战争胜利后，只有部分号家恢复生产，其余的皆因遭受战乱摧残过重，元气大伤，一时难以振兴恢复。此时复业的"中兴源丝织厂"（其前身即"正源兴"客帮号家）只开织机4台，少量地生产一些库缎和妆金库缎，专销给一些来华观光的外国人，对业务发展持观望的态度，年产量只有2300m，产值仅9000多元（旧法币）。后来，虽有部分机台陆续恢复生产，但多是为中兴源丝织厂加工织造的机户。自产自销的小业户，能够复业生产的则为数极少。到1949年4月南京解放前夕为止，全市云锦织造机台仅存150台左右，而能生产的仅有中兴源丝织厂时织时停的4台而已。

After the victory of War of Resistance against Japan in 1945, only some workshops resumed production, while the rest were severely damaged by the war, which made it difficult to revitalize and recover for a while. At this time, ZHONG XING YUAN Silk Weaving Factory（whose predecessor was ZHENG YUAN XING workshop）resumed business, but only started four looms and produced a small amount of Palace satin and Gold Palace satin, which were exclusively sold to some foreign tourists, with an annual output of only 2,300 meters and an output value of only 9,000 yuan（legal tender issued by the National Government）. To the business development, it still took a wait–and–see approach. Although later some loom households resumed production one after another, most of them worked for ZHONG XING YUAN Silk Weaving Factory. Very few self–produced and self–marketing small loom households could resume production. Up to the Nanjing's liberation in April 1949, there were only about 150 Yun brocade weaving looms left in the city, among which only 4 looms in ZHONG XING YUAN Silk Weaving Factory could produce but were often stopped.

四、云锦的新生/The Reborn of Yun Brocade

1949年中华人民共和国成立前夕，许多南京云锦古老的技艺已经失传，许多优秀的传统老纹样也零落散失。中华人民共和国成立后，百业待兴。为了保存这一古老的文化遗产并满足国内外消费者的需求，对南京云锦技艺的抢救、挖掘、整理、研究，已经刻不容缓。在地方政府的大力扶持下，南京云锦很快恢复生产和销售。

Before the People's Republic of China was founded in 1949, many ancient skills of Nanjing Yun Brocade and many excellent traditional old patterns were lost. When the People's Republic of China was founded, all industries were waiting for revival. In order to preserve this ancient cultural heritage and meet the needs of consumers at home and abroad, it is urgent to rescue, excavate, sort out and study Nanjing Yun Brocade craftsmanship. With the strong support of the local government, Nanjing Yun Brocade quickly resumed production and sales.

1953年，第一次"全国工艺美术展览会"在北京举办。这次展览，是中华人民共和国成立后对全国传统工艺美术品的一次大检阅。党和国家领导人先后参观了展览。南京云锦业送展的展品中有一件"大红地加金龙凤祥云妆花缎"，这件彻幅的金龙彩凤妆花缎料气势雄大，富丽辉煌，引起了各界参观者的兴趣和注意，各新闻媒体争相报道，中央及有关专家更是给予热情关注。展览会闭幕后，华东军政委员会文化部根据中华人民共和国文化和旅游部的指示，专门颁布了挖掘、整理、研究南京云锦等民间工艺美术遗产的指示。南京市人民政府文化处（南京市文化局前身）根据指示的精神，结合本地具体情况，确定拯救已濒临艺绝边缘的南京云锦。

In 1953, the first "National Arts and Crafts Exhibition" was held in Beijing. It was a national traditional arts and crafts inspection after the founding of the People's Republic of China. Party and state leaders visited the exhibition successively. There was a piece of gold Zhuanghua Satin of red ground with patterns of dragon, phoenix and auspicious clouds in the exhibition delivered by Nanjing Yun Brocade Industry. This magnificent satin aroused the interest and attention from all walks of life, various news media rushed to report it, and the Central Committee and relevant experts had paid much attention to it. After the exhibition, the Ministry of Culture of the East China Military and Political Commission issued a special instruction to excavate, sort out and study the folk arts and crafts heritage such as Nanjing Yun Brocade according to the instructions of the Ministry of Culture and Tourism of the People's Republic of China. According to the instructions and local conditions, the Department of Culture of Nanjing Municipal People's Government (predecessor of the Bureau of Culture of Nanjing Municipality) decided to save Nanjing Yun Brocade, which was faced up with the crisis that the brocade art gradually declined.

1954年6月，南京市文化局抽调了局美术组的部分美术干部，吸收了云锦技艺水平甚高的张福永（1903～1961年）、吉干臣（1892～1976年）两位老艺人（中华人民共和国成立后仅有4位老艺人健在），组建了"云锦研究工作组"，对南京云锦进行有计划的艺术整理和研究工作。文化局邀请了在宁的我国著名的工艺美术家、美术教育家、工笔花鸟画大师陈之佛（1896～1963年）教授担任名誉组长，负责对云锦研究工作进行指导（图3-13）。文化局美术组组长何燕明兼任云锦研究工作组组长。

In June 1954, the Bureau of Culture of Nanjing Municipality selected some art cadres from the Art Group and invited Zhang Fuyong (1903–1961) and Ji Ganchen (1892–1976), two

图3-13 陈之佛教授（左）与云锦老艺人吉干臣（右）
Professor Chen Zhifo (Left) and Yun Brocade Veteran
Artisan Ji Ganchen (Right)

veteran artisans with high brocade skills（only four veteran artisans were still alive after the founding of the People's Republic of China）, to set up a "Yun Brocade Research Working Group" to carry out the planned artistic collation and research on Nanjing Yun Brocade. The Bureau of Culture also invited Professor Chen Zhifo（1896–1963）, a famous Chinese arts and crafts artist, art educator and master of meticulous flower–and–bird painting in Nanjing, as the honorary team leader to guide the research work of Yun brocade（Figure 3–13）. He Yanming, head of the Art Group, also served as the head of the Yun Brocade Research Working Group.

1957年12月，经过筹备，江苏省政府正式批准成立了"南京云锦研究所"，这是中华人民共和国成立后国家批准的第一个工艺美术专业研究机构（图3-14）。从此，南京云锦研究所作为全国唯一的云锦专业研究机构，承担着云锦传承和保护的历史重任。2004年南京云锦研究所改制成为南京云锦研究所有限公司，2011年底公司完成了股份制改造，更名为南京云锦研究所股份有限公司。多年来，在老所长汪印然和董事长兼总经理王宝林的带领以及全所员工的拼搏开拓下，南京云锦研究所不但得以新生，而且不断发展壮大。

In December 1957, after preparation, the Jiangsu Provincial Government officially approved the establishment of Nanjing Yun Brocade Research Institute, which was the first professional research institution of arts and crafts approved by the state after the founding of the People's

图3-14 南京云锦研究所（南京云锦博物馆）
Nanjing Yun Brocade Research Institute（Nanjing Yun Brocade Museum）

Republic of China (Figure 3-14). Since then, Nanjing Yun Brocade Research Institute, as the only professional research institution of Yun brocade in China, has undertaken the historical responsibility of inheriting and protecting Yun brocade. In 2004, it was restructured into a limited company. At the end of 2011, the company completed the shareholding system reform and changed its name to Nanjing Yun Brocade Research Institute Co., Ltd. Over the years, under the joint efforts of Wang Yinran, the old director, Wang Baolin, the chairman and general manager, and the staff of the whole institute, Nanjing Yun Brocade Research Institute has continuously developed and expanded.

多年来，历届国家领导人和部、省、市级领导都对南京云锦的发展给予了高度的重视和关切，社会各界有识之士和专家学者也都给予积极的支持和帮助，南京云锦的传承一定会有更大的进展。

And successive national leaders and ministerial, provincial and municipal leaders have attached great importance to and concerned about the development of Nanjing Yun Brocade, and people from all walks of life, experts and scholars have also given much support and help, so the Nanjing Yun Brocade will surely get better inheritance.

第二节　云锦的种类与花色/Varieties and Patterns of Yun Brocade

一、云锦的种类/The Varieties of Yun Brocade

在悠久的发展过程中，云锦形成了许许多多的品种。从现在已经掌握的资料来看，云锦大致可分为库缎、库金、库锦、妆花四类。

In the long course of development, many varieties of Yun Brocade had been formed. According to the available data, they can be roughly divided into four types: palace satin, palace gold brocade, palace brocade and Zhuanghua.

（一）库缎/Palace Satin

"库缎"又名"花缎"或"摹本缎"。库缎原是清代御用贡品，以织成后输入内务府的缎匹库而得名。

Palace satin, also known as "jacquard satin" or "Mopen", was a royal tribute in the Qing Dynasty, which was so named because it would be imported into the warehouse of satin of the Internal Affairs Office after it was woven.

"库缎"包括起本色花库缎、地花两色库缎、妆金库缎、金银点库缎和妆彩库缎几种。

Palace satin includes self-tone jacquard palace satin, two-tone jacquard palace satin, gold palace satin, gold and silver palace satin and palace satin embellished with colored velvet, etc.

1. 起本色花库缎/Self-tone Jacquard Palace Satin

起本色花库缎是指单色（经、纬颜色相同）提花库缎、库金、库锦、妆花织物，其特点是在缎地上织出本色的图案花纹。花纹分为"亮花"和"暗花"两种。所谓"亮花"和"暗花"是从织物的外观效果相对而言的。"亮花"的组织是经向缎组织显花，纬向显地纹；从织成效果来看，花纹明显地浮于缎面之上；花部光泽比地部强，故称为"亮花"。"暗花"的组织是纬向显花，经向显地纹；织成效果是花纹明显地凹陷于地部缎纹之下；花部光泽较地部为暗，故称为"暗花"。"亮花"和"暗花"实际是经纬组织变化所产生的两种不同的织成效果。在应用上，一般较粗壮的花纹用亮花表现效果较好；较细致的花纹用暗花表现为宜。

Self-tone jacquard palace satin is the monochromatic fabric of jacquard satin, characterized by the same color of the weft and warp, i.e., weaving patterns in the same color with the satin ground. Patterns are divided into "bright patterns" and "shadow patterns", which are distinguished by their appearance. The satin with bright patterns displays patterns by warp and the ground by weft; as regards to the visual effect, its patterns obviously float on the satin surface; and the luster of the patterns is brighter than that of the ground—hence the name. And for the satin with the shadow patterns, it displays patterns by weft and the ground by warp; the patterns are obviously sunken under the ground, and the luster of the patterns is darker than that of the ground. "Bright patterns" and "shadow patterns" are two different patterns produced by the change of warp and weft fabric weave. In terms of application, generally bold and rough patterns are better displayed with bright patterns, while fine and delicate patterns better in shadow patterns.

2. 地花两色库缎/Two-tone Jacquard Palace Satin

它是两色提花织物，地部为一种颜色，花纹则用另一种颜色的纬线织出。例如，清代江宁织局织造的串枝花缎——"红绿缎"，地部组织为绿色的五枚缎，大红色的纬绒三枚斜纹显花；地花两色相互衬托，花纹非常明显而突出，具有明快而醒目的效果（图3-15）。这种地花两色的库缎，过去官办织局织造较多，民间机坊很少生产，有订货时方才织造。

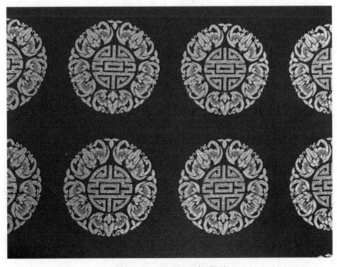

图3-15　地花两色库缎
Two-tone Jacquard Palace Satin

Two-tone jacquard palace satin is a two-color jacquard fabric, whose ground part is of one color, and the patterns are woven with weft threads of another color. The Red and Green Satin, a kind of brocade satin with floral scrolls patterns woven by the Jiangning Official Weaving Bureau in the Qing Dynasty, just belongs to this category. Its ground fabric weave is green five-heddle satin, and three-heddle weft velveteen twill patterns in bright red; the two colors set off each other, and the patterns are very prominent, bright and eye-catching（Figure 3-15）. In the past, two-tone jacquard palace satin was mostly manufactured by the officially-run weaving bureaus, while folk shops rarely wove it unless there was an order.

3. 妆金库缎/Gold Palace Satin

这种库缎，整个缎料的花纹起本色花，所不同者，就是在单位纹样里有局部花纹是用金线装饰的。例如，"五福捧寿""八仙庆寿""二龙捧金寿"等纹样中，五只蝙蝠、八件宝物、两条龙都是起与地部同色的亮花或暗花。而当中的"寿"字则是用金线织出，因而在织成效果上，使图案主题的"寿"字非常明显而突出。这不仅丰富了花纹的层次，而且由于装饰了金线，使整个织物增添了华贵的效果。又如"缠枝牡丹金八宝"妆金库缎，是在满地有规律的暗花缠枝牡丹中，穿插着几个造型不同，并缠有灵活飘带的金色小八宝，既增添了织物的富丽感，又使得规律性很强的缠枝花纹生动活泼起来。此外，如金心莲妆金库缎，整个缠枝莲花是本色起花，而莲心部分装饰了一个金色的小莲蓬，用金虽不多，却起到了富丽别致的装饰效果（图3-16）。

Gold palace satin also has the patterns with the same color of ground, but it is featured that a part of each pattern unit is inlaid with gold threads. For example, those palace satins with traditional patterns, such as "five bats surrounding the Chinese character SHOU" "eight immortals celebrating birthday" and "two dragons encircling Chinese character SHOU", belong to this category. Five bats, eight Daoist Emblems and two dragons are all bright or shadow patterns in the same color as the ground. Among them, the Chinese character SHOU is all woven with gold threads, striking and eye-catching. The inlaid gold threads not only enrich the gradation of patterns, but also add an opulent touch to the whole fabric. On the gold palace satin with patterns of scrolling peonies and Eight Treasures, there are several golden small Eight Treasures of different shapes with fluttering ribbons, interspersed

图3-16　金心莲妆金库缎
The Gold Palace Satin with Golden-hearted
Lotus Patterns

among the shadow-patterned scrolling peonies, which not only increases the luxurious feeling, but also makes the regular patterns of scrolling floral branches vivid and lively. In addition, on the gold palace satin with golden-hearted lotus patterns, the whole scrolling lotuses are in the same color with the ground, while the seedpod of the lotus is woven with gold threads. It does not use many gold threads, but looks splendid and novel (Figure 3-16).

4. 金银点库缎/Gold and Silver Place Satin

金银点库缎的织法与妆金库缎织法相同，只是局部花纹是用金、银两种线装饰。所谓"点"者，就是说妆金妆银仅是极小的一部分。金银点库缎的图案，用"锦群"（又名"天华锦"）格式的图案居多。设计时在单位纹样中，选择间距适度的小型几何花纹或图案式的小朵花，装饰以金线和银线，因而使整件暗花织料增色不少。在织造方法上，妆金库缎、金银点库缎的全部花纹是用通梭织造。而妆金妆银部分的花纹，是用装有金线或银线的小梭进行局部挖花盘织，因此织工们也称其为"挖花库缎"。

Gold and silver place satin is woven in the same way as that of gold palace satin, but its patterns are decorated with gold and silver threads, and only a small part of patterns are decorated. Its patterns mostly follow the format of Tianhua brocade. In each pattern unit, small geometric patterns or small floral patterns are moderately spaced, decorated with gold and silver threads, which adds much luster to the whole shadow-patterned fabric. In terms of the weaving technology, all the patterns of gold palace satin as well as gold and silver place satin are woven with regular shuttles; only a small part is woven in the way of swivel weaving through small shuttles with gold or silver threads.

5. 妆彩库缎/Palace Satin Embellished with Colored Velvet

妆彩库缎是在起明、暗花的库缎上（这里所称的"暗花"是指起本色花的库缎，"明花"是指地花两色的库缎），用彩绒装饰部分花纹。单位纹样以本色暗花为主，或以与地色不同的另一单色明花为主。选择能活跃全局的部分花纹加以彩妆。例如，"凤穿牡丹"的妆彩库缎，不同姿态的飞凤，用与地部同色的暗花表现；穿插于飞凤之中的折枝牡丹，则用多样的彩绒装饰，远看彩花朵朵，近看飞凤随光线的变幻则隐现于花朵之间，素凤彩花，相映成趣。又如，地花两色的库缎，在满地折枝花卉中，穿插着几只飞舞姿态生动的妆彩蝴蝶，整个纹样就显得春意盎然、热闹生动起来。过去这种妆彩的库缎，主体花纹用本色花表现的为多，只在局部花纹上妆一点彩。民间作坊习惯把这个品种叫作"妆夹暗"，因此"妆夹暗"就成了"妆彩库缎"的别名（图3-17）。

Palace satin embellished with colored velvet is woven as follows: on shadow-patterned palace satin (i.e., self-tone jacquard palace satin) or bright-patterned palace satins (i.e., two-tone jacquard palace satin), some patterns are adorned with colored velvet. The pattern unit is composed of all shadow patterns, or its major part is a bright pattern in a different color from that of the ground; only some patterns that can vivify the whole fabric are decorated with colored velvet. Palace satin embellished with colored velvet with phoenix and peony patterns is such an

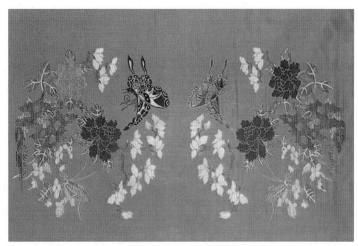

图3-17　妆彩库缎
Palace Satin Embellished with Colored Velvet

example, on which flying phoenixes with different postures are displayed by shadow patterns of the same color as the ground and the peony branches among the flying phoenixes are decorated with various colored velvet. When people look at the satin from a distance, they can see the colorful flowers; while when they look closer, they will see that with the subtle light changes, the flying phoenixes are looming among the flowers, and the phoenix and colorful flowers contrast with each other. Another example is the two–tone palace satin, on which among the floral branches all over the ground, a few butterflies woven with colored velvet are flying, showing a luxuriant and lively spring scene. In the past, on this kind of satin the main pattern was mostly in the same color as the ground, and only a little colored velvet was embellished on a small part of patterns（Figure 3–17）. Folk workshops used to call this "Zhuang Jia'an"（which means "shadowed patterns adorned with colored velvet"）, which was another name of "palace satin embellished with colored velvet".

　　"妆金库缎""金银点库缎""妆彩库缎"都是在明暗花库缎的基础上运用不同的装饰手法变化发展而成的。

　　The above three kinds of palace satins use different weaving techniques on the basis of bright–patterned and shadow–patterned satin.

　　库缎是做衣料用的，民间作坊通常又称其为"袍料"。库缎的花纹设计，用团花居多。生产时，根据成件衣料所需的长度，把花纹设计安排在衣服穿着时显眼的几个主要部位（如前胸、后背、肩部、袖面、下摆）。织成款式完整的成件衣料；制作时只要按式裁剪，即可缝制成衣。而花纹的布列则非常妥适且恰当。这种云锦织品，在丝织工艺上叫作"织成"。它是按照实用物的具体形式和尺寸规格，进行设计和整个图案布局安排的。过去封建帝王的龙袍一般用"织成"的方法织造。过去的"八团花"衣料，也是一种典型的"织成"衣料的设计款式。1911年辛亥革命以后，由于社会生活习尚的改变，人们对衣着装饰的喜爱心理也随之发生很大的变化，袍料上用团花纹样的渐少，"散花"和"折枝花"纹样逐渐流行起来。

Palace satin is used for apparel fabric, usually called as "robe material" in folk workshops. Its patterns are mostly rounded patterns. When it was made into clothes, according to the required length of the ready-made garment, the patterns are designed for several conspicuous main parts of the garment (such as chest, back, shoulders, sleeve and hem). Since the layout of patterns is very fit and appropriate, the woven satin fabrics can be sewn into a complete garment as long as it is cut according to the pattern designs. The way this kind of Yun brocade fabric is manufactured is named "ZHI CHENG (Woven)", a traditional silk weaving technique, which means designing and arranging the layout of patterns according to the specific form and size specifications of the products ready to make. In the past, the dragon robes for emperors were generally woven in this way. The eight-roundel-patterned apparel fabric is a typical design style of "Woven" fabrics. After the Revolution of 1911, due to the change of social customs, people's aesthetics of apparel and accessories also changed greatly. The roundel patterns were rarely seen on the robe, while the patterns of "scattered flowers" and "broken floral branches" gradually became popular.

库缎匹料每件长 7m，幅宽 75cm。花纹单位最多的为"八则"，一般织"四则"花和"六则"花的居多。所谓"则"数，就是指在缎料的幅宽尺寸内，横向并列的单位纹样的数目。"四则"就是横向并列有四个相同的单位纹样。"六则"就是横向并列有六个相同的单位纹样。

Each piece of palace satin is 7m long and 75 cm wide. On a piece of palace satin, there are usually four or six pattern units horizontally within the width size of satin, eight ones at most. "Pattern Unit" is used as the measuring unit to count the number of the pattern units.

（二）库金/Gold Brocade

"库金"又名"织金"，也是因织成以后输入宫廷的"缎匹库"而得名。所谓"库金"，就是织料上的花纹全部用金线织出。也有花纹全部用银线织成的，叫作"库银"。库金、库银属同一个品种，分类上统称"库金"。

"Gold brocade", also known as "Gold-wefted Brocade", was so named because it would be imported into the warehouse of satin of the Internal Affairs Office after it was woven. The so-called "Gold brocade" means that all the patterns are woven with gold threads. There is also brocade with patterns all woven with silver threads, which is called "Silver brocade". Gold brocade and Silver brocade belong to the same variety, which are called "Gold brocade" together in classification.

明、清两代江宁官办织局生产的库金，金、银线都是用真金真银制成，在每匹织料的尾部均织有"××织造真金库金"字牌。说明所用的金、银线材料货真价实（图3-18）。从明、清两代传世的库金锦缎看，由于金线材料考究，虽经过数百年的时间，至今仍是金光灿烂，光彩夺目。

The gold and silver threads used in Gold brocade produced by Jiangning officially-run Weaving Bureau in the Ming and Qing Dynasties were all made of genuine gold and silver. At the tail end of each brocade fabric was woven with the plate with characters as "Gold brocade with

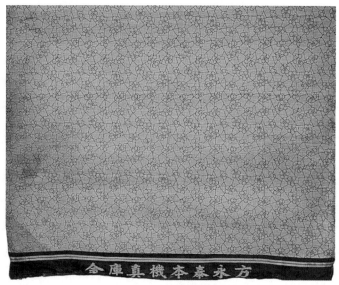

图3-18　方永泰本机库金
Gold Brocade Woven by Fang Yong Tai's Loom

genuine gold woven by ××", which proved that the gold and silver used for threads was genuine (Figure 3-18). From the Gold brocades handed down from the Ming and Qing Dynasties, due to the high-quality gold threads, they are still brilliant and dazzling even after hundreds of years.

设计库金纹样时，要求花满地少，充分利用金线材料，发挥显金效果。传统的库金图案，多采用花纹单位较小的小花纹样。在满金地上，利用地部组织勾画出花纹的轮廓线条来。这种阴纹的轮廓线条，既是图案花纹的具体形象，又是满地金花外的地纹，构思非常巧妙。这种巧妙的设计手法，使价值贵重的真金线在织品的正面得到了充分的利用，达到了最大限度的显金效果。

The pattern designs of Gold brocade require more patterns and less ground, and gold effect is fully displayed by using gold threads. Traditional Gold brocade patterns mostly adopt small-sized patterns with smaller pattern units. On the all-over gold ground, the outlines of the patterns are sketched by the ground fabric weaves. The concave outlines not only display the patterns, but also show the ground, which is an ingenious design. It fully utilizes the valuable genuine gold threads on the fabric, and displays gold effect to the maximum extent.

由于金线制作技术的进步和织造技术的日趋精巧，在清代的云锦织品中，还出现了一些织造技术难度很大的宽幅库金织物，如宽幅的织金陀罗经被。这些特殊用途的宽幅库金云锦，需用特制的宽幅织机，由五六个人同时协作方能织造起来。一件织品，从设计到织成，往往要用数年的时间才能完成。

Due to the progress of gold threads manufacturing technology and the increasingly exquisite weaving craftsmanship, some broadloom Gold brocade fabrics calling for some very difficult weaving techniques were produced in the Qing Dynasty, such as the Broadloom Gold Brocade Dharani Sutra Quilt. These special-purpose broadloom Gold brocades were woven with a special broad loom by five or six craftsmen working together at the same time. It often took several years to finish a broadloom Gold brocade product from design to weaving.

库金主要用于镶滚衣边、帽边、裙边和垫边等。纹样设计通常用十四则的小花纹单位，发展至晚清时期，也采用"七则"较大花纹的设计。明代的库金，多用片金（扁金）织造，光泽效果较好，但不如清代用捻金（圆金）织造的织物质地牢。传统的库金织物纹样有曲水纹锦、冰梅锦、串菊锦（也称洋莲锦）等。七则大花纹的库金织物纹样有缠枝莲锦、八吉龟甲锦、凤莲锦等。

The Gold brocade is mainly used as the edge binding of clothes, hats, skirts and cushions. Its pattern design usually includes fourteen small pattern units. About the late Qing Dynasty, the design of "seven larger pattern units" was adopted. Gold brocade in the Ming Dynasty was mostly woven with flake gold (flat gold). It had a better luster, but its texture fastness was slightly inferior to that of twisted gold (round gold) in the Qing Dynasty. Excellent pattern designs of traditional Gold brocade include Flowing Water, Plum Blossoms on the Ice-cracked Ground, Chrysanthemum Scrolls (also known as Passionflowers) and so on. Pattern designs consisting of seven pattern units include Entangled Lotus Floral Branches, Eight Auspicious Symbols and Tortoiseshell, Phoenix and Lotus Flowers and so on.

（三）库锦/Palace Brocade

云锦中的"库锦"，在民间作坊中分类较为含混，有些属于"库锦"类的织物，被混淆于"妆花"类织物中。因此，首先需要弄清楚这两类织物的异同之处。"库锦"和"妆花"的相同之处是：它们都是采用精练过的熟丝染色后织造，两者都是提花的多彩纬织物。这两类织物的不同之处主要体现在花纹配色和织物背面上，具体如下。

The classification of "Palace brocade" in folk workshops was rather vague, and some belonging to "Palace brocade" were confused with "Zhuanghua" fabrics, resulting in conceptual confusion. Therefore, it is necessary to distinguish the similarities and differences between these two kinds of fabrics. The similarities between "Palace brocade" and " Zhuanghua" are that their raw materials are dyed and woven with degummed silk, and both are multicolored jacquard weft-knit fabrics. The differences between them are mainly reflected in the pattern color matching and the back of fabrics.

（1）花纹配色不同。妆花的花纹配色是用不同颜色的彩绒纬管，对织料上的图案分别作局部的挖花妆彩，配色非常自由，色彩变化丰富。库锦花纹的配色情况就不同了，它是用不同颜色的彩梭通梭织彩，受织造工艺条件的制约，织料上每一段最多只能配织四五种颜色，更换其他颜色必须在花纹单位反复循环时，或织到一定距离时（如"织成"料上花纹变换时），才能陆续更换配色，且每一段上的用色，均不能超过也不能少于规定的用色数量。因此，库锦花纹配色没有妆花自由，色彩变化也不如妆花丰富。

Differences in pattern color matching are as follows. With regards to the Zhuanghua fabrics, the weft shuttle tied with a variety of color velvet performs warp-through and weft-broken swivel weaving part by part on the patterns; its color matching can be conducted freely, subject to no limits. As for the Palace brocade, different general shuttles of colored floss are used to weave colors. Restricted by the weaving process, each section of the fabrics can only be matched with four or five colors at most. Other colors are changed only when the pattern units repeat, or when the fabric is woven to a certain distance (such as when the patterns change on the "woven" fabric), the colors can be changed one after another, and the colors used on each section cannot exceed or be less than the specified number. Therefore, compared to Zhuanghua fabrics, its color matching is less free, and its color change is less rich.

（2）织物背面不同。妆花的配色是用各种颜色的彩色绒纬管对花纹的各个局部作通经断纬的挖花妆彩，因而织品的背面有彩色抛绒（或叫回梭穿绒）。又因配色复杂、彩纬多，织料比较厚重，且厚薄不均匀（妆彩多的花纹部位较厚，不妆彩的地部较薄）。库锦的配色，因采用彩梭通梭织彩，整个彩纬都被平均地织进织料中去，在显花的部位，彩纬露于织料的正面；在不显花的部位，彩纬被织进织料的背面。在织料的地经中，有一根地经是专门用来压住背面彩纬的，形成经纬绞组（作坊术语叫作"扣背"），因此整个织料厚薄均匀，背面光滑服帖。

Differences in the back of fabrics are as follows. With regards to the color matching of Zhuanghua fabrics, since the weft shuttle tied with a variety of color velvet performs warp–through and weft–broken and swivel weaving part by part on the patterns, on the back of the fabric there is colored velvet; because of the complexity of color matching and many colorful weft threads, the woven fabric is thick and uneven in thickness (the pattern parts with more colors are thicker, and the ground parts without patterns are thinner). As for the color matching of Palace brocade, the whole color weft threads are evenly woven into the fabric because they apply the regular shuttles with colored floss. In the figured part, the color weft threads are exposed on the front of the fabric, while in the unfigured part, the color weft threads are woven into the back of the fabric. Among the ground warp threads of the fabric, one piece of ground warp thread is specially used to press the colored weft on the back, forming a warp and weft intertwining group (called "back buckle" by some weaving workshops), so the whole fabric is even in thickness and smooth on the back.

云锦中属于库锦类的织物有许多品种，如二色金库锦、彩花库锦、抹梭妆花、抹梭金宝地、芙蓉妆等。

There are many kinds of Palace brocade fabrics in Yun brocade, and the following varieties are common, including Gold and Silver Palace Brocade, Multicolor Palace Brocade, Shuttling Zhuanghua, Shuttling Jinbaodi, Hisbiscus Zhuanghua and so on.

1. 二色金库锦/Gold and Silver Palace Brocade

二色金库锦属小花纹单位的织锦。花纹全部用金、银两种线织出，一般是以金线为主，小部分花纹用银线装饰。图案花纹以几何形纹样和小朵花为多。二色金库锦的具体用途是作为服饰和实用物的缘边装饰，如镶滚衣边、帽边、垫边等。花纹单位有十四则、二十一则、二十八则几种。

Gold and Silver Palace Brocade belongs to the brocade with small pattern units. All patterns are woven with gold and silver threads; gold threads are mainly used, while silver threads are used to embellish a small part of the pattern. Most of the patterns are geometric patterns and small floral patterns. The Gold and Silver Palace Brocade is used as mainly used as the edge binding of apparel, accessories and some other practical objects, such as that of clothes, hats or cushions. The pattern designs of Gold and Silver Palace Brocade usually include fourteen, twenty-one and twenty-eight small pattern units.

2. 彩花库锦/Multicolor Palace Brocade

彩花库锦也简称为"彩库锦"，也是一种小花纹单位的织锦。彩花库锦的花纹除用金线织造外，还用各种颜色的彩绒装饰极小部分的花纹（如一朵小花或某一花纹的局部）。彩花部分用通梭织彩、分缎换色的办法，全件织料上的各段彩花只用几种不同的颜色循环，或者全部花纹只用一金、一彩两色长跑梭织造。彩花库锦用色虽不多，但织品效果却甚为妩媚精巧、美观悦目。花纹单位常用的有十四则、二十一则、二十八则几种。彩花库锦除用作衣物的镶边装饰外，也可用于制作囊袋、锦匣、枕垫和装帧装潢（图3-19）。

Multicolor Palace Brocade, also known as "Multicolor Floral-patterned Palace Brocade", also belongs to the brocade with small pattern units. Its patterns are woven with gold threads as well as colored velvet which is used to decorate a very small part of the pattern (such as a small flower or a part of a certain pattern) with various colors of. Its multicolor patterns are woven by general shuttles and colors are changed by sections. Multicolor patterns in each section of the whole fabric

图3-19　彩花库锦
Multicolor Palace Brocade

only use several colors, or all patterns are woven by a long-range shuttle with gold threads and another long-range shuttle with color threads. Although there are not many colors in Multicolor Palace Brocade, the fabric looks very charming and exquisite. The pattern designs of Multicolor Palace Brocade usually include fourteen, twenty-one and twenty-eight small pattern units. Besides being used as the edge binding of clothes, it can also be used to make bags, brocade boxes, pillows, bookbinding and calligraphy or painting mounting（Figure 3-19）.

3. 抹梭妆花/Shuttling Zhuanghua

抹梭妆花是一种大花纹的彩锦，又可分为加金线装饰花纹和不加金线装饰花纹两种。所谓"抹梭"，就是指整个织物花纹的配色是用通梭织彩，在显花的部位，彩纬呈现在织料的正面；在不显花的部位，彩纬织进织料的背面（有地经与其形成经纬的交组）。抹梭妆花的配色，一般在同一段上最多织四五种颜色。因为是通梭织彩，所以在同一段上循环并列的花纹单位图案与配色也都完全相同；在织物的背面，呈现着一段段的横向彩条。为使图案的配色丰富而富于变化，采用分段换色等办法。但在整体配色中，必须有一两种颜色作为统一整体色彩的统色。例如，一匹缠枝花纹样的织料，常用深、浅两种绿色作为枝梗和藤叶的配色，花朵的彩妆，则可分段换色。这样，从整体效果看，色彩既有变化，又有统一。这种织造统色的彩梭，在织造术语上叫作"长跑梭"；分段换色的彩梭，叫作"短跑梭"。抹梭妆花的织造速度虽比挖梭妆花要快一些，但用木机手工生产，产量有限。熟练的织工，一天最多只能织60cm。

Shuttling Zhuanghua is a kind of multicolor brocade with large patterns, which has two forms, with gold（patterns decorated with gold threads）and without gold. "Shuttling" means that the color matching of its patterns is realized by shuttling; in the figured part, the colored wefts appear on the front side of the fabric, while in the unfigured part, the colored wefts are woven into the back side of the fabric（there is an intersection of ground warp and weft）. In the same section, four or five colors at most are woven, and the pattern units and color matching of pattern designs repeated in the same section are also the same; on the back of the fabric, there are sections of horizontal color bars. In order to obtain a rich and changeable color matching, colors change by sections. However, in the overall color matching, there must be one or two colors as the unified colors. For example, a fabric with the pattern of entangled floral branches often uses deep and light green as the matching colors of branches and leaves, while the color of flowers can be changed by sections. In this way, as a whole, the colors of the patterns look changeful but harmonious and unified. The shuttle used for weaving unified colors is called "long-range shuttle" while the shuttle used to change colors by sections "short-range shuttle". Although Shuttling Zhuanghua is woven faster than Swivel-shuttle Zhuanghua, the output is still limited by manual production with a wooden loom. Skilled weavers can only weave about 60 cm a day.

抹梭妆花除织造匹料外，清末以后民间作坊曾生产过一些不加金的织锦台毯、织锦靠垫和提包等实用品。图案有花卉、人物、博古、龙凤、琴棋书画、亭台楼阁等。整个织品只用固定的四五种颜色装饰全部花纹，色彩虽不多，效果倒很雅致。过去云锦行业叫它"洋庄货"，因其主要销售对象是来华旅游的外国人和归国观光的海外侨胞，于20世纪30年代曾经风靡一时。

After the end of Qing Dynasty, folk workshops produced some practical brocade products without gold such as brocade blankets, brocade cushions and bags, with patterns of flowers, figures, antiques, "Dragon and Phoenix" "Lyre-playing, Chess, Calligraphy and Painting" "Pavilions, Terraces and Towers", etc. The whole fabric only used four or five colors to decorate all patterns, not rich in color but quite elegant. In the past, this kind of products was called "foreign goods" in Yun brocade industry, because its target consumers were foreign tourists and overseas Chinese returning to homeland for sightseeing. It was all the rage and very popular in 1930s.

4. 抹梭金宝地/Shuttling Jinbaodi

"抹梭金宝地"的织造方法、配色技术与"抹梭妆花"完全一样。所不同者，"抹梭妆花"是缎地，"抹梭金宝地"是用捻金线织满地。前者是在缎地上织彩花，后者是在满金地上织彩花。织品的背面均光平且呈现彩条，花纹轮廓均用片金绞边（也称"金包边"）。从织品效果看，抹梭金宝地更显得辉煌华丽。

The weaving methods and color matching techniques of Shuttling Jinbaodi are the same as those of Shuttling Zhuanghua. Their difference is that Shuttling Zhuanghua, multicolor patterns are woven on the satin ground, while on Shuttling Jinbaodi, multicolor patterns are woven on the all-over gold ground. The back of both fabrics is light and flat, displaying multicolor strips, and their pattern outlines are both embellished with flaked gold (also known as "gold leno-edge"). Judging from the fabric effect, Shuttling Jinbaodi is more brilliant and gorgeous.

在云锦的妆花织物中，还有一种叫作"金宝地"的织物。这是我国传统丝织品中南京特有的品种。

Among the Zhuanghua fabrics in Yun Brocade, there is also a unique traditional variety of Nanjing called "Jinbaodi".

"金宝地"是用圆金线（即捻金线）织满地。在金地上织出五彩缤纷、金彩交辉的花纹来，整个织品极为辉煌而富丽。

Jinbaodi has the all-over ground woven with round gold threads (or twisted gold threads) and colorful and golden patterns on the gold ground, which makes the whole fabric extremely brilliant.

"金宝地"除用满金作地外，图案花纹的织金妆彩方法，与妆花缎完全一样。花用扁金包边、挖花妆彩，但花纹的装饰手法比妆花缎更为丰富多样。织料的主体花纹和妆花缎一样，用多层次的色彩表现（运用"色晕"的方法表现）。陪衬的宾花，妆彩方法要比妆花缎丰富得多。有用色晕表现的，也有用单色表现的，还有用扁金、银线装饰的，或金、银线并用装饰的。彩花用扁金包边，金花、银花则用彩绒包边。这些丰富多彩的装饰手法，在整件织料的花纹装饰上，运用得非常巧妙而妥帖。如用金或银，不但量多，而且有着多种多样的表现手法和装饰效果。用圆金织的满金地，光泽含蓄而沉着，花纹绞边的扁金和用扁金、银线装饰的花纹，光泽闪耀而明亮，形成了不同的金色色调和不同的光泽效果。大量的金或银巧妙地运用在五彩斑斓的彩花中，又起着调和与统一全局色彩的作用，使整个织品金彩交辉，具有一种辉煌而华贵的气派。

Except the all-over gold ground, the pattern weaving technique of Jinbaodi is exactly the same as that of "Zhuanghua satin". As Zhuanghua Satin, its patterns are also edged with flaked gold and woven by swivel weaving techniques, but its pattern decoration techniques are more diverse; the main patterns are also expressed by multi-level colors (using the method of "halo color"), but the auxiliary patterns are woven by more diversified weaving techniques, including halo color, and monochrome, or embellishing with flat gold or silver threads or with both gold and silver threads, edging colored patterns with flat gold, and edging golden and silver patterns with colored velvet. These decorative techniques are used skillfully and properly in the pattern decoration. In terms of gold or silver using, it not only uses a large amount, but also has various ways of expression and shows diverse decorative effects. The all-over gold ground woven with round gold threads has subtle and calm luster, and the flat gold used as edges and the patterns decorated with flat gold and silver threads have shining and bright brilliance, forming different golden tones and different luster effects. A large amount of gold or silver is skillfully used in colorful patterns to harmonize and unify the overall color, which makes the whole fabric golden and colorful, with a very brilliant and luxurious style.

金宝地是从元代金锦演进而来。从织品的设计和成品的效果来看，可以说它是"库金"和"妆花"这两个品种结合运用的产物。故宫保存的很多清代流传下来的金宝地实物，无论是花纹、色彩还是织造技艺，都反映了我国传统丝织技艺的高超水平，如图3-20所示。

Jinbaodi evolved from Gold brocade in Yuan Dynasty. From the design and finished products, it can be said that it is a combination of "Palace brocade" and "Zhuanghua satin". Many Jinbaodi

图3-20 百花金宝地
Jinbaodi with Patterns of Hundreds of Flowers

fabrics handed down from Qing Dynasty preserved in the Forbidden City, whether in pattern, colors or weaving techniques, reflect the superb traditional silk weaving craftsmanship in China, as shown in Figure 3–20.

5. 芙蓉妆/Hibiscus Zhuanghua

芙蓉妆是一种配色比较简单的大花纹织锦，在色彩的表现方法上与"妆花"有着显著的不同。芙蓉妆的花纹不用片金绞边（用片金织出花纹的轮廓），也不用深浅不同的几重色彩来表现花纹的层次，整个纹样只用几种不同的色块来表现。花纹的形状以空出地部线条来显现。在花与花、花与叶之间，各以不同的单色表现配色的变化。如花卉的枝梗、花萼、叶芽等，以一色绿或者两种颜色的绿作为整体的统色，用长跑梭织造；花朵则用短跑梭分段换彩，以变换各段花头的配色。如一段织红头、一段织黄头、一段织金头……可以随意变化安排。芙蓉妆整个织品的配色，虽不如妆花和金宝地那样复杂、丰富，但它另有一种艳而不繁、单纯而明快的效果。由于它配色单纯，施用的彩纬不多，因此整个织料质地薄而平整。由于在过去用这种妆彩方法织造的织锦的图案多以芙蓉花为主题，所以作坊中习惯称其为"芙蓉妆"（图3–21）。后来凡是应用这种彩妆方法织造的"锦"，花纹虽不用芙蓉作为主题，但仍习惯把它叫作"芙蓉妆"。

图3-21 芙蓉妆
Hibiscus Zhuanghua

Hibiscus Zhuanghua is a large-patterned brocade with simple color matching, significantly different from Zhuanghua Brocade in color expression. The pattern of Hibiscus Zhuanghua does not need to be edged with gold (embellishing the pattern outlines with flaked gold), nor does it need several colors with different shades to display the gradation of patterns. The patterns are only expressed by several different color blocks, and the pattern shape is revealed by vacant ground lines. Between flowers and flowers, flowers and leaves, different monochromatic colors show the color changes. For example, branches, calyxes, leaf buds, etc. are woven with long-range shuttles with one-tone or two-tone green as the overall color; flowers are woven by short-range shuttles to change colors by sections and change the color matching of each corolla, i.e. corollas can be woven in red, yellow, gold or other colors in each section, arranged at will. The color matching of Hibiscus Zhuanghua fabrics is not as complex and rich as that of Zhuanghua Brocade and Jinbaodi, but colorful, simple and bright. Because of its simple color matching and few colored weft, its texture is thin and flat. In the past, this brocade woven in this method was so named by the folk workshops because the fabric was mostly hibiscus-patterned（Figure 3-21）. Later, although "hibiscus" was not always used as the pattern theme, people still used to call it "Hibiscus Zhuanghua".

（四）妆花/Zhuanghua

妆花是云锦中织造工艺最为复杂的品种，也是最具南京地方特色和代表性的提花丝织品种。

Zhuanghua is the most complicated variety of Yun brocade in weaving techniques, and also the most representative jacquard silk variety with Nanjing local characteristics.

"妆花"是织造技法的总名词。始见于明代的《天水冰山录》，该书记载，查抄严嵩家时抄出的大批织物中，有很多"妆花"名目的丝织物，如妆花缎、妆花绸、妆花罗、妆花纱、妆花䌷、妆花绢、妆花锦等。妆花织物有加织金线的，也有不加织金线的。

Zhuanghua, originally as one of the founding weaving techniques, first appeared in *Tianshui Iceberg Records* in the Ming Dynasty. The book records that among the large number of fabrics confiscated from Yan Song's home, there were many silk fabrics with the name of Zhuanghua, such as Zhuanghua satin, Zhuanghua silk, Zhuanghua half-cross leno, Zhuanghua yarn, Zhuanghua bourette silk, Zhuanghua lustre, Zhuanghua brocade, and so on. Zhuanghua silk fabrics were woven with or without gold threads.

妆花织物的特点是用色多，色彩变化丰富。在织造方法上，是用各种绕有不同颜色的彩绒纬纡管，对织料上的花纹做局部的盘织妆彩，配色极度自由，没有任何限制。图案的主体花纹，通常是用两个层次或三个层次的色彩来表现，部分花纹则用单色表现。一件妆花织物，花纹配色可多达十几种乃至二三十种颜色。妆花的用色虽然多，但均能处理得繁而不乱、统一和谐，使织物上的花纹获得生动而优美的艺术效果。

Zhuanghua silk fabrics are featured with many matching colors and rich color changes. With regards to the weaving method, the weft shuttles tied with a variety of color velvet perform warp-through and weft-broken and swivel weaving part by part on the patterns of the fabric. The color matching is extremely free without any restrictions. The main patterns are usually expressed in two or three levels of color, while some minor patterns are displayed in monochrome. For a Zhuanghua silk fabric, the matching colors can be as many as ten or even twenty or thirty. Although many colors are used in Zhuanghua silk fabrics, they look complicated but harmonious, which makes the patterns vivid and beautiful.

　　妆花织物失传的品种很多，目前留传的仅有"妆花缎"一种。妆花缎是在缎地上织出五彩缤纷的彩色花纹，色彩丰富，配色多样。以四则花纹单位的妆花缎匹料为例，在同一织物幅面上横向并列有四个连续花纹单位，每个花纹单位的图案完全一样，由于运用了通经断纬、挖花盘织的妆彩技法，四个单位的花纹配色可以不相雷同（当然这里还须注意到相互间的色彩关系）。充分显示了妆花织物妆彩工艺的长处和特点。若织造"织成"形式的妆花织物（如龙袍、桌围、靠垫、装饰挂屏等），则更能发挥这种配色的特长，如图3-22所示。

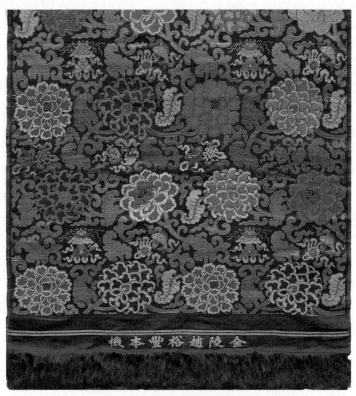

图3-22　四则八吉牡丹莲妆花缎
Zhuanghua Satin of Four Pattern Units with Patterns of
Eight Auspicious Symbols, Peony and Lotus

Most of the Zhuanghua varieties have been lost, and only "Zhuanghua satin" has been passed down. On Zhuanghua satin, colorful patterns are woven on satin ground, rich in color and diverse in color matching. Taking the Zhuanghua satin with four pattern units as an example, there are four continuous pattern units horizontally juxtaposed on the same fabric surface, and the patterns of each unit are exactly the same. Because of the application of the warp–through and weft–broken and swivel weaving technique, the matching colors of the four pattern units can be different (the color harmony should also be noted), which fully shows the advantages and characteristics of Zhuanghua technique (Figure 3–22). Zhuanghua silk fabrics in the form of "ZHI CHENG (Woven)" (such as dragon robes, table skirts, cushions, decorative hanging screens) can better reflect the uniqueness of this color matching scheme.

二、云锦的花色/The Patterns of Yun Brocade

1. 云锦的配色/The Color Matching of Yun Brocade

我国彩锦的生产，有悠久的历史。在元代盛行用金风气的影响下，明、清两代的织锦既重视配色，又考究用金，两种装饰方法兼收并蓄，形成金彩并重的锦缎装饰新风貌。

The production of colorful brocade in China has a long history. Under the influence of the prevailing trend of using gold in the Yuan Dynasty, the brocade of the Ming and Qing Dynasties not only attached importance to color matching, but also used gold fastidiously. The two decorative methods were integrated, forming a new style of brocade decoration with both gold and color.

在色彩感情上，我国的传统向来喜爱温暖、明快、鲜艳和强烈的积极色，不喜欢弱色和多次的间色。因此，青、红、黄、绿、紫、白、黑成为我国装饰的主要色彩，云锦的用色也继承了这一民族装饰用色的传统。在元代用金风气的影响下，结合御用服饰和宫廷装饰的具体实用要求，形成了云锦独特的用色规律和装饰特色。

Chinese traditionally prefer warm, bright and strong colors than gloomy secondary and tertiary colors, so cyan, red, yellow, green, purple, white and black are the main colors for decoration in China. The colors of Yun brocade also inherits this tradition. Under the influence of the fashion of using gold in the Yuan Dynasty, combined with the specific practical requirements of royal costumes and palace decoration, it formed its own color rules and decorative characteristics.

云锦图案的配色，主调鲜明强烈，具有一种庄重、典丽、明快、轩昂的气势。这种配色手法与我国宫殿建筑的彩绘装饰艺术是一脉相承的。就妆花缎织物的地色而言，浅色很少应用，除黄色是特用的地色，多采用大红、深蓝、宝蓝、墨绿等深色作地色；而主体花纹的配色，也多用红、蓝、绿、紫、古铜、藏驼等深色装饰。"色晕"和色彩调和处理手法的运用，使得深色地上的重彩花，获得了良好的艺术效果，形成了整体配色的庄重、典丽的主调，与宫廷里辉煌豪华和庄严肃穆的气氛非常协调，并与封建帝王的黄色御服起着对比衬托的效果。

The color matching of brocade patterns has a distinct tone, solemn, beautiful, bright and grand. This color matching technique comes down in line with the art of decorative painting of palace architectures in China. As far as the ground color of Zhuanghua satin fabric is concerned, light colors are rarely used. Except yellow, which is a specific royal ground color, dark colors such as bright red, dark blue, sapphire blue and myrtle green are mostly used as the ground color, while dark colors such as red, blue, green, purple, antique bronze and camel mostly as the matching colors of main patterns. Due to the skill of "color halo" and color harmony, the heavy color patterns on the dark ground are well displayed and the overall color matching looks solemn and beautiful, which is in harmony with the brilliant, luxurious and solemn atmosphere in the palace, and sets off the yellow imperial robes.

在云锦图案的配色中，还大量使用金、银（金线、片金，银线、片银）这两种光泽色。

金、银两种色可以与任何色彩相调和。妆花织物中的花纹全部用片金绞边，部分花纹还用金线、银线装饰。金银在设色对比强烈的云锦图案中，不仅起着调和和统一全局色彩的作用，而且使整个织物增添了辉煌的富丽感，使之更加绚丽悦目。这种金彩交辉、富丽辉煌的色彩装饰效果，是云锦特有的艺术特色。

In the color matching of brocade patterns, gold and silver (gold thread, flake gold, silver thread and flake silver) are also widely used. Because gold and silver are two colors, they can be harmonized with any color. All the patterns in the Zhuanghua Satin fabric are twisted with flake gold, and some patterns are decorated with gold threads and silver threads. In the brocade patterns with strong color contrast, gold and silver not only play a role in harmonizing and unifying the overall color, but also add brilliant richness to the whole fabric, making it more gorgeous, which is the unique artistic feature of Yun brocade.

色晕的运用也是云锦配色的一大特色。色晕有两种方法，里深外浅的（如最里层晕色为大红，中曾晕色为浅红，外层晕色为粉红），叫作"正晕"（图3-23）；外深里浅的，叫作"反晕"（艺人也叫它"反绞"）。正晕、反晕的运用，是根据纹样"显妆"的要求和整体色彩变化的需要来决定。通常采用的是正晕，如图3-24所示。

The color matching of Yun brocade also adopts the decorative method of "halo color". There are two kinds of color halo. The one with deep inside and shallow outside is called "positive-halo" (Figure 3-23) (for example, the innermost halo color is bright red, the middle halo color is light red,

图3-23 宝蓝地加金"普天同庆"妆花缎
Gold Zhuanghua Satin of Sapphire Blue Ground
with Patterns of Peonies and Chrysanthemums
（Symbolizing "a Universal Rejoicing"）

图3-24 三色晕
Tricolor Halo

and the outer is pink); the other with deep outside and shallow inside is called "reversed-halo"(also called "reversed-twisting"). Whether to use positive-halo or reversed-halo is determined by the effect of setting off patterns and overall color changes. Positive-halo is more often used. Figure 3-24 is such an example.

2. 云锦的图案/Patterns of Yun Brocade

云锦的图案取材广泛，花纹内容极为丰富，其中绝大部分是取自现实生活中人们所熟悉的自然素材，也有部分富有浪漫主义色彩的纹样。

The patterns of brocade come from a wide range of sources, and the content of patterns is extremely rich. Most of them are taken from natural materials realistic and familiar to people, while some of them with strong romantic elements.

明清时期，南京是士子云集的文化之都，南京云锦无论从审美情趣，还是色彩纹样方面，无疑都深受南京士大夫们的贵族文化的影响。云锦的装饰纹样几乎"图必有意，意必吉祥"，汇集了优秀的中国吉祥文化的精粹。

Since Nanjing in the Ming and Qing Dynasties was a capital of culture where scholars gathered, Nanjing Yun Brocade was undoubtedly deeply influenced by the noble culture of Nanjing scholar-bureaucrat in terms of aesthetic taste and color patterns.The decorative patterns of Yun brocade were almost auspicious patterns, a distillation of excellent Chinese auspicious culture.

从素材上看，南京云锦图案囊括了动物、植物、佛道、乐器、文房四宝、人物、传统吉祥内容等写实的或几何形式的纹样。从文化内容上看，有期望功名富贵、升官发财的，有祈祝好运的，有期望平安和气的，有宣扬封建伦理纲常的，有宣扬封建社会人际关系的，还有期望子孙繁衍、聪明富贵的。总之，有来自民间文化的牡丹、蝙蝠、鱼（富余）、石榴（多子多孙）、桃（长寿）等内容，也有反映皇室文化的龙、凤、麒麟、江涯、海水、万寿等内容，且二者有机地揉合在一起，形成了以皇室文化为主的具有独特文化色彩的图案。其表现形式有单独纹样、二方连续、四方连续或三者的结合，生动活泼又富于变化，如图3-25和图3-26所示。

From the perspective of themes, Nanjing Yun Brocade patterns include realistic or geometric patterns such as animals, plants, Buddhism and Taoism symbols, musical instruments, Four Treasures of the Study (writing brush, ink stick, ink slab, paper), figures and traditional auspicious items. From the cultural perspective, Yun brocade patterns reflect diversified traditional cultural concepts or good wishes, such as the pursuit of fame, fortune and high official rank, the praying for good luck, the hope for peace and harmony, the yearn for more offspring and the wealth and wisdom of children, the promotion of feudal ethics or interpersonal relationships in feudal society. Among them, some come from folk culture, such as peony, bat, fish (symbolizing "a surplus of wealth"), pomegranate (symbolizing "many offspring"), peach (symbolizing "longevity"), while others from royal culture, like dragon, phoenix, Kylin, river, seawater and Chinese character SHOU (longevity). The two are often combined and form patterns with strong cultural colors

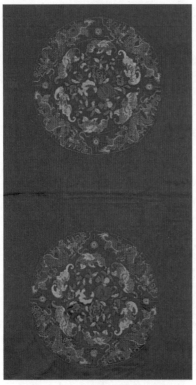

图 3-25　五福捧寿四龙纹锦
Brocade with Patterns of Four Dragons
and Five Bats（Symbolizing Blessing）
Surrounding the Peach（Symbolizing
Longevity）

图 3-26　云八宝八仙纹绸
Silk with Patterns of Clouds, Eight Daoist
Emblems and Eight Treasures

mainly based on royal culture, expressed in forms of single patterns, or continuous patterns in double directions, continuous patterns in four directions or the combination of the three, lively and full of changes, as shown in Figure 3-25 and Figure 3-26.

　　由于南京云锦在明清两代主要是御用品，所以图案中多为龙、凤、麒麟、狮、江涯、海水、牡丹、寿桃等。其中引人注目的是象征天子、帝王权力的龙的表现形式，有正龙、团龙、盘龙、升龙、降龙、卧龙、行龙、飞龙、侧面龙、七显龙、出海龙、入海龙、戏珠龙等多姿多彩的不同形态，以及与此相配的日、月、星辰、山、龙、华虫、宗彝、藻、火、粉米、黼、黻等十二章纹（图3-27）。用单独纹样大云龙、大凤莲等整匹云锦面料装饰宫廷，显出豪华而又威严的气派。另外，云纹也有百种之多，如四合云、如意云、和合云、七巧云、蚕茧云、骨朵云、海潮云、大勾云、小勾云、行云、卧云等（图3-28）。

Because Nanjing Yun Brocade was mainly woven for the emperors in the Ming and Qing Dynasties, dragon, phoenix, Kylin, lion, river, seawater, peony, peach, etc., are often seen in its patterns. Among them, the dragon, which represents the deified power of "Son of Heaven" and emperors, is the most striking pattern. There are different forms of dragons, such as sitting dragon (with its head facing forward）, round dragon, crouching dragon, rising dragon（with its head at the top and tail at the bottom）, descending dragon（with its head at the bottom and tail at the top）,

图3-27 清红地云龙纹吉服袍
The Red-grounded Robe with Patterns of Clouds
and Dragons

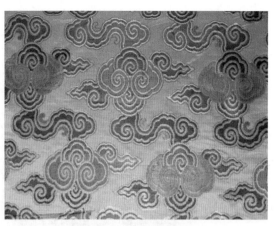

图3-28 四合如意云纹
Sihe Ruyi Cloud Pattern

lying dragon, roaming dragon, flying dragon, dragon in profile, out-of-the-sea dragon, into-the-sea dragon and dragon playing with a pearl. The dragon patterns are matched with Twelve Symbols, i.e. the sun, the moon, the constellation, the mountain, the dragon, the Huachong（the flowery bird or the pheasant）, Zongyi（the sacrificial vessels）, the waterweed, the flame, the Fenmi（a bowl of rice）, the Fu（黼, the axe head）, and the Fu［黻, the confronted Chinese character JI（己）］（Figure 3-27）.The whole brocade fabrics with separate patterns such as "cloud and dragon" and "phoenix and lotus" are often used for palace decoration, showing a luxurious and dignified air. In addition, there are hundreds of cloud patterns, such as Sihe Cloud, Ruyi Cloud, Hehe Cloud, Qiqiao Cloud, Silkworm Cocoon Cloud, Bud Cloud, Tide Cloud, Large Hook Clouds, Small Hook Clouds, Drifting Cloud, Lying Cloud and so on（Figure 3-28）.

祥瑞寓意的图案在宫廷官府更为常见，如灵仙祝寿、麒吐玉书、百子庆寿（图3-29）、八吉凤莲、事事如意、吉庆双余、六合同春、瓜瓞绵绵、百果丰硕、五谷丰登、喜相逢、并蒂莲、太平有象、福寿三多（石榴子多、佛手福多、寿桃寿多）等。中国人认为"家和万事兴"，主张血亲同居一室，向往大家庭的和睦团结、安康、幸福生活，如用鹌鹑、菊花组合图案，九只鹦鹉比喻"九世"，菊花既是君子，又借"菊"与"居"谐音，组成"九世同居"，是中华民族传统的吉祥兆瑞象征。

Patterns with auspicious cultural connotations are more commonly seen on the Yun brocade fabrics for palace court or officials, such as "LING XIAN ZHU SHOU（ganoderma lucidum celebrating longevity）", "QI TU YU SHU（a Kylin presenting jade books）", "BAI ZI QING SHOU（a hundred boys celebrating birthday, as shown in Figure 3-29）, "BA JI FENG LIAN（eight auspicious symbols with phoenix and lotus）", "SHI SHI RU YI（a Ruyi and two persimmons）, "JI QING SHUANG YU（chime stone and double fishes）", "LIU HE TONG CHUN（deer, cranes, flowers and pine trees）, "GUA DIE MIAN MIAN（scrolling melons and vines）", "BAI GUO FENG SHUO（plentiful fruits）", "WU GU FENG DENG（a bumper grain harvest）", "XI

图 3-29　百子庆寿妆花缎
Zhuanghua Satin with Hundred Boys Celebrating
Birthday Pattern

XIANG FENG（double fishes）", "BING DI LIAN（Twin lotus flowers on one stalk）", "TAI PING YOU XIANG（an elephant carrying a vase）", "FU SHOU SAN DUO（pomegranate seeds, bergamots and peaches）and so on. Chinese people believe that a peaceful family will prosper, and advocate that people of the same blood should live together, so they also prefer the traditional auspicious pattern combined by quails and chrysanthemums, which means "nine generations living together" and symbolizes the harmony, unity and happiness of the extended family.

　　云锦图案严谨庄重，纹样变化概括性很强，配色浓艳而大胆，设计上非常注重花纹造型和章法的处理。一幅纹样，不管采用的素材有多少，经过设计艺人的匠心处理后，均能达到繁而不乱、疏而不凋、层次分明、主题突出的艺术效果。

　　The patterns of Yun brocade are rigorous and solemn, with strong variability, rich and bold color matching, and great attention is paid to the pattern style and composition in design. No matter how many imageries are used in patterns, the whole fabric can achieve the artistic effect that is complicated but not messy, sparse but not boring, with distinct levels and prominent theme after being designed with ingenuity by craftsmen.

　　云锦图案的花纹造型，很多虽以生活为基础，但它并不拘泥于生活和自然形象。很多花纹的造型，是根据主题表现和装饰变化的需要，采用"添加"的手法，使之装饰化或寓意化。例如，牡丹花头的造型，多是运用均齐对称的形式表现。为了在均齐对称中求得生动的变化，往往在花心部位，安放一件"八宝"，如如意头、双犀角或双古钱，以代替烦琐的花心和花蕊（图3-30）。这不仅不影响花头的庄重造型和形象的完整，相反地却起到"平中求奇"的艺术效果，打破了过分统一的单调感，增添了纹样的装饰意趣，并赋予纹样以吉祥的内容含义。又如，在柿子纹中装饰一个如意头，则赋予"事事如意"的吉祥含义。这种不拘泥于写实的造型手法，极富浪漫主义色彩。

Although many patterns of Yun brocade are based on daily life, they are not rigidly attached to daily life and natural images. Many patterns are designed according to the theme expression and decorative changes, using the technique of "adding" to make them decorative or symbolic. For example, the pattern of peony's corolla is mostly expressed in uniform and symmetrical form. In order to realize vivid changes in uniform symmetry, one of "Eight Treasures", such as Ruyi Head, double rhinoceros' horns or double ancient coins, is often placed in the center of flower to replace the stamens and pistils（Figure 3-30）. It does

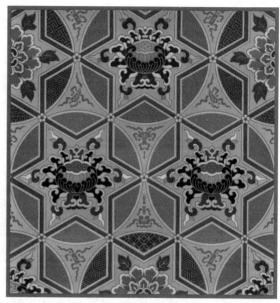

图 3-30　花心部位安放 "如意头" 的云锦图案
The Brocade Pattern with Ruyi Head in the Center of Flower

not disturb the solemn and image integrity of the corolla, but achieves the artistic effect of "seeking novelty in the plain" by breaking the excessively unified monotonous, adding the decorative interest, and endowing the pattern with auspicious connotations. Another example is to decorate the persimmon pattern with a Ruyi head, which provides the auspicious meaning of "everything goes well". This modeling technique, not rigidly attached to realism, is full of romantic elements.

云锦图案的布局，非常讲究呼应与气势。一幅纹样中，如有两种主题花（如缠枝牡丹莲，牡丹与莲花均为主题花），除造型上处理成一尖瓣、一团瓣形成对比的变化外，两种花头的方向，总是安排成相对的呼应关系。在 "主花" 与 "宾花" 的关系上，除了有花形大小的对比外，图案的内容含义和花纹的布局安排总是有着内在的思想联系和形式上的呼应关系（图 3-31）。

The layout of the Yun brocade patterns is very particular about coordination. In patterns on the whole fabric, if there are two kinds of theme flowers（such as in patterns of entangled floral branches of peony and lotus, both peony and lotus are theme flowers）, the directions of the two flower corollas always correspond with each other. In the relationship between "main flowers" and "auxiliary flowers", besides the comparison of shape and size, the connotations and the layout of patterns always have certain correlation in both culture connotation and form（Figure 3-31）.

云锦图案花纹的造型变化极其丰富多彩。以常用的牡丹与莲花来说，造型的变化就有上百种之多，并各有其生动而优美的姿态。

The pattern designs of Yun brocade are extremely rich and varied. With the commonly used "peony" and "lotus" patterns, there are hundreds of pattern variants and each has its vivid and beautiful posture.

图 3-31　缠枝牡丹莲云锦图案
The Brocade Pattern with Entangled Floral Branches of Peony and Lotus

　　云锦图案在继承我国丝织品传统图案基础上，形成了自己的一套装饰格局和表现形式。根据织品的具体使用情况，设计上可分为两大类："织成"和"料"。

　　On the basis of the traditional patterns of silk fabrics in China, Yun brocade has formed its own decorative layout and expression forms in terms of patterns. According to the specific use of fabrics, the fabrics can be divided into two categories: " ZHI CHENG（Woven）" and "Material" .

　　"织成"是按实用品种的具体形式和规格要求设计织造的织料，如过去的龙衣、蟒袍、佛像、经被、伞盖等御用服料和宫廷用的特殊织料，以及生活中实用的台毯、椅垫、桌围、靠垫、提包、挂屏和固定款式的"八团"衣料等。除了龙衣、蟒袍、佛像、经被等有着固定纹饰内容和特定表现形式的特种织品外，其他"织成"形式的实用品基本上都是应用适合纹样的图案格局。料子织好后，只要按式缝缀和做一些装潢加工（如台毯镶上须边，提包配以框架，垫子镶以边饰等），即可成为一件文饰安排妥适、形式完美的实用品或装饰品。组成料在纹样的设计安排上，是颇费匠心的。很多特殊的"织花"（如龙袍、彩织佛像等）在制作和织造工艺上，技术非常复杂，一件织品需要耗费大量的工时，如图 3-32 所示。

　　"ZHI CHENG（Woven）" refers to the fabrics designed and woven according to the specific form and specification requirements of practical varieties. For example, in the past, imperial apparel fabrics for dragon robe, ceremonial robe embroidered with pythons, etc., fabrics for palace decoration such as Buddhist embroidery works, sutra quilt and umbrella cover, practical articles for daily use such as blanket, chair cushion, table skirt, cushion, bag, hanging screen, as well as the fixed-style eight-roundel-patterned apparel fabric, etc., all belong to this category. Except for special fabrics such as dragon robe, ceremonial robe embroidered with pythons, Buddhist

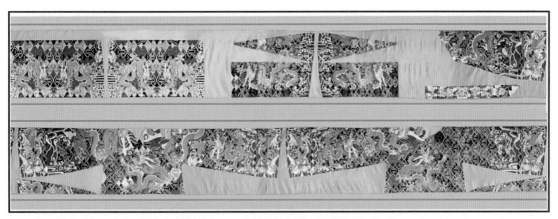

图 3-32　云锦妆花缎直身龙袍织成料
The Woven Zhuanghua Satin Fabrics for a Straight Dragon Robe

embroidery works and sutra quilts, which have fixed patterns and forms, other practical products in the form of "Woven" basically apply the layout suitable for patterns. After the material is woven, it will become a practical product or ornament with proper pattern arrangement and perfect form, as long as it is sewn according to the style or only needs to be added with some decorations（For instance, the blankets need to be inlaid with fringe edge, the bags to be matched with frames, and cushions to be inlaid with edges）. The "Woven" materials are quite ingenious in the design and arrangement of patterns. The process of weaving "Woven" fabrics with patterns（such as dragon robes, Buddhist embroidery works）is very complicated in manufacturing and weaving technology, and a piece of fabric needs a lot of working hours, as shown in Figure 3-32.

　　"料"是云锦实用品和装饰品的加工用料，是云锦织品的主要生产形式。云锦料的图案花纹，都是应用四方连续纹样组成的。只要设计出一个花纹单位，即可通过织造的工艺手段，使它反复循环，无限连续。根据织品需要的长度，织成成件的料。料花纹单位的大小，是根据不同的织物品种和不同的实用要求决定的。一般来讲，妆花缎、金宝地的花纹单位较大。妆花缎常用的有四则花纹单位和二则花纹单位；做大件装饰用的，则用独花（即一则花纹单位）纹样。金宝地的门幅宽度较妆花缎窄，常用的是二则花纹单位，根据实用需要也有独花纹样的。这种大花纹单位的设计，其他地区生产的锦缎中是少有的，是云锦中妆花和金宝地纹样设计上的一个明显特点。库锦是镶衣边和帽边用的，花纹单位较小。常用的花纹单位规格有十四则、二十一则和二十八则几种，也有织七则花纹单位的库金。花纹大小适度，适于做装潢用料。库缎常用的四则花纹单位较多，也有织二则花纹单位的。

　　"Material" refers to those fabrics that need to be processed for practical products and decorations of Yun brocade，and it is the main production form of Yun brocade fabric. The patterns of Yun brocade Material are all in the form of continuous patterns in four directions. As long as a pattern unit is designed, it can be repeated infinitely through the weaving process. Material is woven into pieces according to the required length of fabric. The size of Material pattern unit is determined by different fabric varieties and practical requirements. Generally speaking, the pattern units of Zhuanghua satin and Jinbaodi are larger. As for Zhuanghua satin, four pattern units and two pattern units are commonly used; if it is used for the decoration of large objects, the single pattern unit is used. In terms of Jinbaodi, since it has a smaller width than Zhuanghua satin, the two pattern units are often used, and single pattern unit is also adopted according to practical needs. The large pattern unit is an obvious feature of the pattern design of Zhuanghua and Jinbaodi, rare in brocade fabrics produced in other areas. Palace brocade is used for clothes and hat edges, so its pattern unit is smaller. Fourteen, twenty-one and twenty-eight pattern units are commonly used, and sometimes seven pattern units are also used on Gold brocade. Its patterns are of moderate size, suitable for decoration. As for Palace satin, it often uses four pattern units, and sometimes two pattern units.

云锦图案设计，是紧密结合生产制作条件并考虑其章法布局的。它一方面受生产条件局限性的制约，另一方面又利用了生产制作的可能条件，创造了省工取巧、降低工本的可能性，以收到事半功倍、省工省时的经济效果。例如，四方连续、八方接章的图案构成方法，在明清两代云锦图案的设计中，应用得非常广泛而纯熟。这种经济而省工的设计和制作技巧，值得我们继续效仿。

The composition and layout of Yun brocade patterns is closely combined with the production conditions. On the one hand, it is restricted by the limitations of production conditions; on the other hand, it makes use of the production conditions and creates all possibilities to reduce cost, save both labor and time, so as to yield twice with the half efforts. For instance, continuous patterns in four directions and eight sides were widely and skillfully used in the design of Yun brocade patterns in the Ming and Qing Dynasties. This economical and labor-saving design and manufacturing techniques are worth emulating today.

第三节　云锦的传承/The Inheritance of Yun Brocade

一、云锦的应用/The Application of Yun Brocade

保护非物质文化遗产是为了传承和发扬，促进人类社会的可持续发展。南京云锦的传承和保护工作已经取得了可喜的成绩，同样，云锦的利用和发扬也取得了很大的进展。

The purpose of protecting intangible cultural heritages is to inherit and carry forward and promote the sustainable development of human society. Lots of achievements have been made in the inheritance and protection of Nanjing Yun Brocade, as well as in the utilization and development of brocade.

为满足现代生活、内外销和旅游等方面的各种需要，各种新品种应运而生，如云锦台毯、靠垫、枕套、提包、家具装饰和包装、装裱及围巾、领带、服装等（图3-33）。特别是云锦领带，图案采用团寿纹样，周围如意祥云缭绕，寓意吉祥、如意、富贵、长寿；色彩有

图3-33　云锦工艺品（包和领带）
Handicraft Articles Decorated with Yun Brocade（Bags and Ties）

红、黄、蓝、黑、绿、紫等各样鲜艳的颜色，其中遍布金线，光彩夺目，适合各年龄段的男士佩戴。

In order to meet the needs of modern life, domestic and foreign sales and tourism, various new Yun brocade products have been created, such as table cloth, cushion, pillowcase, handbag, furniture decoration, mounting materials, scarf, tie, apparel（Figure 3-33）. Especially the Yun brocade tie, the patterns of Chinese character SHOU were surrounded by auspicious clouds, symbolizing auspiciousness, wealth and longevity. There are red, yellow, blue, black, green, purple and other bright colors, inlaid with glittering golden threads, suitable for men of all ages.

传统的云锦服饰以其独特的个性、丰富的内涵和不俗的表现给人们以典雅华贵的美感，具有很高的美学价值和观赏价值，是一种文化内涵深厚的高档产品。从2003年春节联欢晚会主持人身着云锦服饰惊艳登台，到连续多届参加"南京世界历史文化名城博览会"的各国市长们穿着云锦服饰亮相世界，云锦服饰早已深入人心。新一代云锦人不断努力将云锦中的传统元素运用在现代服饰中，赋予其时代色彩和新意，致力于开发出缎、绸、纱、罗等不同组织结构的面料，以适用于春、夏、秋、冬四季穿着的服饰及其配套产品，真正使云锦服饰面料走出国门，面向世界。

Traditional brocade costumes with the unique characteristics, rich connotation and distinguished design give people elegant and luxurious aesthetic feeling, which have high aesthetic and ornamental value, and are high-grade handicrafts with profound cultural connotation. From the hosts of *the Spring Festival Gala* in 2003 to the mayors from various countries who participated in successive sessions of *World Historical and Cultural City Expo* held in Nanjing, brocade costumes have long been deeply rooted among people. The new generation of Yun brocade craftsmen constantly strive to apply the traditional elements to modern costumes, endow them with the characteristics and innovations of the times, devote themselves to developing fabrics with different fabric weaves such as satin, silk, yarn and half-cross leno which are suitable for costumes and accessories worn all the year round, and truly bring Yun brocade costumes go abroad.

二、云锦的创新/The Innovation of Yun Brocade

云锦技艺的保护与传承是同云锦的创新与发展相同步的。1959年，为庆祝中华人民共和国成立十周年，人民大会堂江苏厅的沙发面料和其他装饰织物、大型摄影画册《中国》的封面、外交部迎宾馆高级沙发面料都采用了在继承传统的基础上织造出来的带有创新成分的新云锦。

The protection and inheritance of Yun brocade techniques keep in step with the innovation and development. In 1959, to celebrate the 10th anniversary of the founding of the People's Republic of China, the sofa fabrics and other decorative fabrics in Jiangsu Hall of the Great Hall of the People in Beijing, the cover of the large-scale photography album *China*, and the high-grade sofa fabrics of the Guest House of the Ministry of Foreign Affairs all adopted new Yun brocade with innovative elements based on the inheritance of tradition.

在传统手工操作的基础上，为寻求云锦的现代生产技术的突破，南京云锦研究所与东南大学、浙江大学合作开发了"云锦CAD辅助设计系统"，实现了云锦设计、意匠过程的计算机化；使用电脑挑花，实现了挑花结本的半自动化，使传统的云锦织造前道工序摆脱了落后的手工操作状态；运用电脑提花与手工挖花盘织相结合，实现了部分云锦产品的半机械化生产。

In order to seek a breakthrough in modern production technology of Yun brocade on the basis of traditional manual operation, Nanjing Yun Brocade Research Institute cooperated with Southeast University and Zhejiang University to develop "Yun Brocade CAD System" to design and draft with computer. The semi-automation of cross-stitch knots is realized by using computer, which makes those traditional working procedures before brocade weaving break through the backward manual operation. The semi-mechanized production of some brocade crafts has also been realized by combining computer jacquard with manual swivel weaving.

结合现代审美观点，云锦人还研发了具有浓郁民族文化特点的京剧脸谱、十二生肖，根据陈之佛国画设计编织的《松鹤延年》，为书法爱好者设计织造的名家书法，如赵朴初的《知恩报恩》、刘浚川的《澹泊明志》，以及设计了广大群众喜爱的"梅兰竹菊"四君子图等一批云锦新产品。这些产品满足了各级厅堂、办公室和中西居室等的装饰要求。

Combined with modern aesthetics, craftsmen have also developed various new products with various patterns, including some patterns with rich Chinese cultural features, such as Peking Opera facial makeups, twelve Chinese zodiac signs; some patterns from well-known painting or calligraphy works, such as Chen Zhifo's traditional Chinese painting *SONG HE YAN NIAN* (pines and cranes, symbolizing longevity in Chinese), Zhao Puchu's calligraphy work *ZHI EN BAO EN* (being conscious of a kindness and trying to repay it), Liu Junchuan's calligraphy work *DAN BO MING ZHI* (showing high ideals by simple living), and the well-loved painting of four plants with noble quality (plum blossom, orchid, bamboo and chrysanthemum). These brocade products are suitable for decorating halls, offices and rooms either in Chinese or Western style.

三、云锦的保护和传承/The Protection and Inheritance of Yun Brocade

南京云锦2006年被国务院列为我国首批非物质文化遗产名录，2009年被联合国教科文组织列入人类非物质文化遗产代表作名录。

Nanjing Yun Brocade was included into the first list of national intangible cultural heritages by the State Council in 2006, and was included into the Representative List of Intangible Cultural Heritage of Humanity by UNESCO in 2009.

云锦织造工艺和技巧是云锦这一非物质文化遗产的核心内容。保护与传承文化遗产，就必须保持好它们传统的原生态状况，并在此基础上发扬光大。

Weaving craftsmanship and skills are the core of Yun brocade as an intangible cultural heritage. To protect and inherit cultural heritages, we must maintain their original traditions and then carry them forward.

○ 第四章
锦的织造技艺
Brocade Weaving Craftsmanship

一、图案设计与配色/Pattern Design and Color Matching

1. 图案设计/Pattern Design

锦的图案设计既继承了传统的现实主义思想，又富有浪漫主义的色彩。

The pattern designs of brocade not only inherit the traditional realism, but also are full of romantic color.

蜀锦的主流纹样一是寓意吉祥的传统纹样，如折枝如意、凤穿牡丹等；二是几何图案纹的旋转、重叠、拼合、团叠，如八答晕锦、六答晕锦等。

There are two main types of pattern designs in Shu brocade. The first type is traditional auspicious patterns. The second type is the rotation, overlapping and splicing of geometric patterns, such as Badayun and Liudayun patterns.

宋锦在图案风格上以变化几何为骨架，如龟背、四答晕、六答晕、八答晕等，内填自然花卉、吉祥如意纹等，配以和谐的地色，略加对比色彩的主花，使之艳而不俗，古朴高雅。既具有唐宋以来的传统风格特色，又与元、明时期流行的光彩夺目的织金锦、妆花缎等品种有着明显的区别，更符合贵族和士大夫阶层崇尚优雅秀美的艺术品位。

In terms of pattern style, Song brocade takes changing geometric patterns as the skeleton, such as tortoiseshell, Sidayun, Liudayun, Badayun, etc., filled with flowers or auspicious patterns, matched with harmonious ground color and main flowers of slightly contrasting colors, making the brocade simple and elegant. The brocade has the traditional style since the Tang and Song Dynasties, but is obviously different from the dazzling gold brocade and Zhuanghua satin popular in the Yuan and Ming Dynasties, which is more consistent with the artistic taste of aristocrats and scholar-officials advocating elegance and grace.

云锦的纹样图案，反映了人们祈求幸福的思想观念性，表达了以中国吉祥文化为核心主题的设计思想，即"权、福、禄、寿、喜、财"六字要素。

The patterns of Yun brocade reflect people's yearning for happiness, and express the core theme of Chinese auspicious culture, that is, the six elements: power, blessing, prosperity, longevity, happiness and wealth.

古代织锦一般用象征、寓意、比拟、表号、谐音等方法来表达图意内容。

In the ancient brocade, symbolism, implication, personification, symbolic tokens, homophones and other methods are often used to express the meaning of the picture.

（1）象征。就是根据某些动植物的生态、形状、色彩、功用等特点来表现某种特定的思想，如石榴内多籽实，象征多子多孙；牡丹花型丰满娇艳、富丽华贵，象征富贵；松树树龄极长，有松寿万年之说，古代又称鹤寿千年，故织造松、鹤图以象征长寿。

Symbolism is an artistic style using the forms, shapes, colors, or functions of some animals and plants images to express a certain idea. For example, pomegranate with many seeds, symbolizes many sons and grandsons; plump, delicate and showy peony blossom, symbolizes wealth and prosperity; pine trees and cranes, which were believed to live for thousands of years in ancient times, symbolize longevity.

（2）寓意。借某种纹样题材原有的特定含义，寄寓吉利的思想，如装饰纹样以桃子寓意长寿，如图4-1所示。

Implication refers to conveying auspicious ideas and thoughts through the original meaning of a certain pattern. For example, the peach patterns symbolize longevity, as shown in Figure 4-1.

（3）比拟。赋予某些题材拟人化的性格。比如，梅花孤高挺秀，耐寒抗雪；松树高大挺劲；竹子虚心有节。故将松、竹、梅比喻"岁寒三友"。又如，将梅花、竹子和幽香沁人的兰花以及素雅高洁的菊花合称"四君子"；用莲花比喻出淤泥而不染，如图4-2所示。

Personification, refers to endowing some objects with human beings' qualities or characters. For example, plum blossoms are thought to be noble, unsullied, and resistant to cold and snow; pine trees lofty and strong; bamboo modest and with integrity. Therefore, they are compared to "the Three Companions of Winter". Plum blossom, bamboo, orchid and chrysanthemum are called "Four Gentlemen". Lotus is compared to a beauty rising unsullied from mud, as shown in Figure 4-2.

（4）表号。例如，以八仙手中所拿器物作为八仙的表号纹样，即铁拐李的葫芦、汉钟离的扇子、张果老的渔鼓、何仙姑的荷花荷叶、蓝采和的花篮、吕洞宾的宝剑、韩湘子的横笛、曹国舅的阴阳板，这八种器物称为"暗八仙"。藏传佛教中的八种器物，即宝瓶、宝盖、双鱼、莲花、右旋螺、吉祥结、尊胜幢、法轮，称为八吉祥，如图4-3所示。

Symbolic tokens, refer to using the symbols as decorative patterns. For instance, the magical objects or instruments held by the Eight Immortals are often used to represent the Eight Immortals. That is, Li Tieguai's gourd, Han Zhongli's fan, Zhang Guolao's fishing drum, He Xiangu's lotus flower, Lan Caihe's flower basket, Lv Dongbin's sword, Han Xiangzi's flute and Cao Guojiu's yin and yang jade tablet, which are called "Covert Eight Immortals". Eight instruments or signs in

Tibetan Buddhism, such as the Treasure Vase, the Precious Parasol, the Two Golden Fish, the Lotus, the Conch Shell, the Knot of Eternity, the Victory Banner and the Wheel of Dharma, are called "the Eight Auspicious Symbols", as shown in Figure 4–3.

（5）谐音。以装饰纹样题材的名称组合成同音词来表达吉祥含义。例如，玉兰、海棠、牡丹谐音为"玉堂富贵"；灵芝、水仙配以寿山石谐音为"灵仙祝寿"；水塘里金鱼数尾、水草飘拂的图案谐音为"金玉满堂"等。

Homophones, refer to combining of the names of decorative patterns and employing their homophonic Chinese characters to express auspicious meanings. For example, magnolia, crabapple and peony are combined as "YU TANG FU GUI（may your noble house be blessed with wealth and honor）"; ganoderma lucidum, Chinese narcissus and Shoushan Stone are combined as "LING XIAN ZHU SHOU（fairy bestows birthday greetings）"; several goldfish in the pond with algae

图4-1　吉庆双鱼蟠桃妆花缎
Zhuanghua Satin with Patterns of the Double–fish and Flat Peach

图4-2　环藤莲花宋锦　　　　　图4-3　蓝地彩织团龙宋锦（八吉祥）
Song Brocade with Patterns of Floral Branch　　　Song Brocade of Blue Ground with Pattern of
Scrolls and Lotus　　　　　　　　　　Winding Dragons（the Eight Auspicious Symbols）

and grass have the auspicious meaning of "JIN YU MAN TANG (affluent life full of gold and jades)".

2. 配色/Color Matching

古代织锦的用色丰富多彩，但要调和处理得庄重和鲜丽就不容易了。锦的图案一般是连续的，如牡丹花，在一匹料子上几百上千朵牡丹，要做到朵朵颜色有别是非常困难的，这就要求设计人员在配色上下功夫。能够做到整幅妆花，众色纷成，交相辉映，但又和谐统一，在斑斑耀采中含有稳重沉静因素，就是非常成功的设计。

The colors of brocade are rich and colorful, but it is not easy to reconcile all the matching colors to achieve the solemn, elegant and beautiful effect. Brocade patterns are generally continuous. For instance, on a bolt of brocade material, there might be hundreds of peonies patterns, so it is very difficult to make each flower in different colors, which requires designers to put in a lot of efforts on color matching. If a brocade with many complicated multicolor patterns looks harmonious and unified as a whole, it can be regarded as a very successful design.

锦的图案配色，很多是根据纹样的特定需要，运用浪漫主义的手法来处理的。云纹常用红、蓝、绿三种色彩来装饰，并以浅红、浅蓝、浅绿三色作外晕，或以白色作外晕，增加其色彩节奏的美感。蓝色的云、绿色的云是违背生活真实的，但古代诗词中有"碧云天"和"一朵妖红翠欲流"这样的描绘，千百年来脍炙人口，并没有人说它不真实，相反却交口赞誉其传神。

The color matching of brocade patterns is often achieved by some romantic techniques based on the specific needs of patterns. For example, cloud patterns are often decorated with red, blue and green, with light red, light blue and light green or only white color as external halo, to increase the aesthetic feeling of the color rhythm. Though blue or green clouds do not exist in the real world, there are some poetic descriptions like "bluish clouds and sky" in classical Chinese poems popular for thousands of years, which have been praised as vivid, rather than untrue.

锦图案上云纹的配色，也是相同的道理。它不仅丰富了视觉上色彩的变化，而且符合人们对神话境界和祥云瑞气的想象和寄托。如生活中的莲花，有红色、粉色和白色，但织锦图案中多用蓝灰和紫灰来表现，这主要是受佛教艺术的影响，宗教艺术常常要求图案效果庄严、沉静和恬然。

The same is true of the color matching of clouds patterns on brocade. It not only enriches the visual color changes, but also conforms to people's imagination of the fairyland with auspicious clouds. Another example is the lotus flower, which should be red, pink or white in the real world. However, the lotus flowers as brocade patterns are mostly expressed in bluish gray and purplish gray. It is mainly influenced by Buddhist art which requires solemnity, silence and serenity.

锦的传统配色方法是极为丰富多彩的，它们都是来自历代织锦艺人长期辛勤的艺术实践，是无数织锦艺人汗水和智慧的结晶，如图4-4～图4-6所示。

The traditional color schemes of brocade are extremely varied, which come from the arduous

图4-4　纹样配色设计稿1
Pattern Color Scheme 1

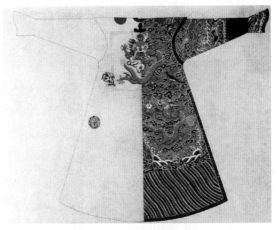

图4-5　纹样配色设计稿2
Pattern Color Scheme 2

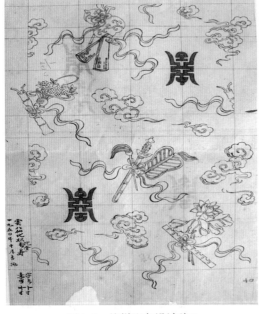

图4-6　纹样配色设计稿3
Pattern Color Scheme 3

artistic practice of countless brocade craftsmen in the past dynasties and are the crystallization of their sweat and wisdom, as shown in Figures 4-4 to 4-6.

二、组织设计和意匠图绘制/The Design of Fabric Weaves and the Drawing of Pattern Grid

1. 组织设计/The Design of Fabric Weave

在纹样设计中，还有一项重要的设计，即组织设计。所谓组织设计，就是织物组织结构的编结程序。云锦，特别是妆花，以七枚缎和八枚缎为主。

The design of fabric weave is one of the important parts of pattern design. The design of fabric weave refers to the knitting procedure of fabric weave. Yun brocades, especially Zhuanghua fabrics, are mainly seven-heddle satain and eight-heddle satain.

当设计人员将纹样、色彩、组织规格等设计好之后，就要根据小样填绘意匠图。意匠纸是特制的，上面有纵横小格，纵横小格代表经纬纱线，小格的纵横比例代表经纬密度。其规格有"八之八"至"八之三十二"，共计25种，根据织物的经纬密度而选择使用。如设计出的织物经纬密度分别为20根/cm×20根/cm，那么意匠纸就可选择使用"八之八"。意匠图是根据纹样的轮廓用铅笔放大，按照织物组织的结构用水粉颜料画成的，意匠图上各种颜色代表着不同的织物组织。

After finishing the design of patterns, colors, fabric weave specifications, etc., designers will fill in the pattern grid according to the hand samples. The design paper is specially made, with vertical and horizontal grids on it. These grids represent warps and wefts, and the aspect ratio of each grid the density of warps and wefts. There are 25 specifications of design paper, from "8 × 8 (eight vertical grids and eight horizontal grids)" to "8 × 32 (eight horizontal grids and 32 vertical grids)", which are selected according to the density of warps and wefts. For instance, if the warp and weft density of the fabric to design is 20 threads/cm × 20 threads/cm, the "8 × 8" design paper would be selected. After designing the patterns, according to the characteristics of the fabric weaves, the patterns are enlarged proportionally and painted with gouache on a piece of checkered paper, which is called pattern grid. Various colors on the pattern grid represent different fabric weaves.

2. 意匠图绘制/The Drawing Pattern Grid

绘制意匠图是一项细致复杂的工艺。绘制意匠图的步骤一般为：第一步，根据品种规格及机台装造，计算并选用适当规格的意匠纸；第二步，将设计好的纹样放大描绘到意匠纸上；第三步，按设计纹样的设色，分成若干场次，在意匠纸放大了的纹样轮廓线范围内，用对应的该场次意匠先勾轮廓再平涂，如图4-7所示。

It is a meticulous and complicated process to draw a pattern grid. The steps of drawing a pattern grid are as follows. Firstly, according to the variety, specifications and samples, the designers count the required vertical and horizontal grid number and select the design paper of appropriate specifications. Secondly, depict the magnified pattern on the design paper. Thirdly, according to the set matching colors of the designed pattern, draw the pattern outline and then fill in colors into the grids inside the outline parts by parts, as shown in Figure 4-7.

根据不同的纹样采用各自相宜的涂法。云锦妆花一般为缎地，可自由勾边；花卉图案讲求流畅，转折自然，忌生硬僵直。宋锦几何图案则讲求严谨规矩。

Different patterns should be delineated in different ways. For example, Zhuanghua fabrics, generally of satin ground, can be freely delineated. When delineating floral patterns on Zhuanghua fabrics, designers should make the pattern lines smooth and natural, and try to avoid being stiff and rigid. As for delineating geometric patterns on Song brocade, designers should pay more attention to the preciseness.

清代卫杰的《蚕桑萃编》中，对挑花纹样的设计制作过程有较详细的记录："取花样，须用五道纸张。第一道，自己想出时新者，画出为式。第二道。昭式画好。第三道，择画工好样式，并四镶安置玲珑者，套画一张。第四道，用底张粘放花样，大小合适。第

图4-7　意匠图
A Pattern Grid

五道，用薄亮细纸，将花样描画干净，然后打横顺格式，用铅粉调清凉水，使笔全抹一遍，为免纸光伤眼。用红绿洋膏子色，记明号码，方好挑取。"

Extract and Compilation of Sericulture, written by Wei Jie in the Qing Dynasty, recorded in detail the design process of cross-stitch patterns as follows. "To form the pattern, five pieces of paper should be used. On the first piece of paper, the designer draws the samples of new patterns that he has come up with. On the second piece of paper, the designer draws the patterns according to the sample. On the third piece of paper, the designer draws the patterns meticulously with tools, such as rulers and trisquares, frames it and copies another one. On the fourth piece of paper, stick the copied pattern in appropriate size. On the fifth piece of paper, which should be thin, bright and fine, the designers first depict the patterns clearly; next draw vertical and horizontal grids; then mix the lead powder with cool water, and apply the mixture with a writing brush on the paper all over again, so as to avoid paper light hurting eyes; finally fill in the red and green pigment, whose shade number should be clearly marked in advance, on the grids."

三、挑花结本/Cross-stitch Work for Forming Decorative Patterns

挑花结本在锦的织造中的作用原理简要地说，就是用丝线（俗称"脚子线"）作经线，用棉线（俗称"耳子线"）作纬线，对照绘制好的意匠图，经线对应意匠图上的纵格，纬线对应意匠图上的横格，挑制成花纹样板（图4-8）。上机织造时，使每根脚子线与织机上的每一根经丝一一通过织机大纤相连接，织造时，通过耳子线提起应该起花的部分，织入彩纬或金、银线，由此美丽的锦就织出来了。

Simply put, the working principle of cross-stitch work for forming decorative patterns in brocade weaving is as follows: silk threads (commonly known as "foot threads") as warps and cotton threads (commonly known as "ear threads") as wefts, a brocade craftsman corresponds the warp threads to the vertical grids, and the weft threads to the horizontal grids according to the drawn pattern grid, and then do cross-stitch work to make the pattern sample (Figure 4-8). When weaving on the loom, the craftsman connects each foot thread with each warp thread on the loom one by one through the harness cord of the loom, next pulls up the part that needs to weave patterns through the ear threads, and then weaves colored weft or gold and silver threads into it. The beautiful brocade is woven in this way.

织锦的挑花结本有三个工艺，称为挑花、倒花、拼花。其中挑花是基本工艺，倒花和拼花是辅助工艺，视情况需要加以

图4-8　根据意匠纸挑花
Doing Cross-stitch Work According to the Pattern Grid

运用。一般来说，任何一个纹样，只用挑花工艺就能编制出供上机织造的花本。但在有些情况下，比如大型的单独纹样，经线数达到1800根甚至2400根时，一架绷子容纳不下（挑花绷子为了操作方便，一般宽为90cm，最多容纳900根脚子线），则需分开挑花，而后运用拼花工艺，拼成一个完整的花本。有些对称型和连续型的纹样，为了节省工时，可以只挑一个基本单位，再用倒花工艺，复制出其他单元，然后采用拼花工艺拼制出完整纹样的花本。倒花工艺还用于复制花本，以供数台织机同时生产或替换已损坏严重的旧花本。

There are three techniques for cross-stitch work for forming decorative patterns, which are called cross-stitch, pattern draft copying and pattern draft connecting. Cross-stitch is an elementary and necessary technique, while pattern draft copying and pattern draft connection are auxiliary techniques, applied only as needed. Generally, any pattern draft used for weaving on the jacquard machine, can be realized through the cross-stitch work. However, in some cases, such as large individual patterns, when the number of warp threads reaches 1,800 or even 2,400, a shed can't accommodate it (for the convenience of operation, the tambour is generally 90cm wide and can accommodate up to 900 foot threads), it is necessary to do cross-stitch work separately, and then use the pattern draft connecting technique to make a complete pattern draft. For some symmetrical and continuous patterns, in order to save working hours, only one pattern unit can be selected first, and then other units can be copied by the pattern draft copying technique, and then the pattern draft of complete patterns can be connected by the pattern draft connecting technique. The pattern draft copying technique is also used to duplicate the pattern draft, so that several looms can produce or replace the old pattern drafts which had been seriously damaged at the same time.

1. 挑花/Cross-stitch Technique

挑花前先将耳子线制成约2m的长度，每8根为一束，两端拼齐后对折成为16根，在距折连顶端3cm处打一个结，形成一个套圈（俗称耳子），然后穿在缰绳上。脚子线的长度及根数需根据不同品种机台的纤线数、纹样长度及色彩场数来确定。耳子线的根数约为意匠图上的横格数与色彩场数的乘积，脚子线的根数与意匠图上的纵格数相等。

Before doing cross-stitch work, make the ear threads about 2 meters long, and tie up every 8 threads in a bundle. After both ends are put together, fold it into 16 pieces in half, tie a knot 3cm away from the top of the folded connection, and form a loop (commonly known as "ear"), and then draw it in the draw string. The length and number of foot threads should be determined according to the number of fiber threads, pattern length and color sections of different looms. The number of ear threads roughly equals to the product of the number of horizontal grids and the number of color sections on the pattern grid, while that of foot threads equals to the number of vertical grids.

把准备好的意匠图、脚子线、耳子线、明纤（代表色彩场次的丝线，如果意匠图上有6种颜色，脚子线旁就要配上6根明纤，分别代表6种颜色）等按一定的要求装上挑花绷子，每一个绷子再备上若干竹制的挑花钩。

Then, prepare the pattern grid, foot threads, ear threads and silk threads representing the color sections (if there are six colors on the pattern grid, six silk threads should be matched with the foot threads, representing six colors respectively), install them on the tambour according to certain requirements, and prepare several bamboo cross-stitch hooks for each tambour.

挑花时用右手执钩，按顺序挑起一根代表色彩场次的明纤，然后根据明纤代表的颜色，依次自右向左看意匠图上同一色彩的纬线起止位置，对照相应的脚子线起止位置，将其范围内的脚子线分段全部挑起，再取左侧备用耳子线一根，钩在挑花钩子上，抽出钩子，把耳子线引入。依明纤标记的色彩场次顺序将本梭各场次挑引完耳子线后，再挑下一梭。如果一梭内有五种颜色，就要挑引五次才算完成一梭。挑完意匠图上的一大格，即八梭后将所引入的耳子线集成一束在尾端打结，以备后面拼花或出现差错时核对，如图4-9和图4-10所示。挑制完成后，将花本下绷进行梳洗整理。

Hold the hook with one's right hand, pick one silk thread representing color section. According to the exact color represented by the silk thread, look at the starting and ending positions of weft threads of the same color on the pattern grid from right to left, then pick all the foot threads in the range between the starting and ending positions of the corresponding foot threads. Pick a spare ear thread on the left side on the hook, then pull out the hook, and draw in the ear thread. According to the sequences of color sections marked by the silk threads representing the color sections, pick the ear threads of this shuttle, and then pick the next shuttle. If there are five color sections in a shuttle, it is necessary to pick five times before completing a shuttle. After picking a large grid on the pattern grid, that is, after eight shuttles, bundle up the ear threads that have been drawn in and tie a knot at the end, so as to prepare for the pattern draft connecting work or have a check when errors occur, as shown in Figures 4-9 and 4-10. After the cross-stitch work is completed, take down the pattern draft, and tidy it up.

2. 倒花/Pattern Draft Copying Technique

倒花就是根据已有的花本复制出另一本花。倒花工艺的基本原理，简而言之就是用祖本

图4-9 挑花结本细节
Details of the Cross-stitch Work for Forming Decorative Patterns

图4-10 挑花结本
The Cross-stitch Work for Forming Decorative Patterns

（已经挑好的花本）脚子线同白本（尚未编结的脚子线）脚子线——兜连，拴系成X形；倒制时，祖本一根耳子线被拽提后，一根浮在耳子线上的脚子线也被提起，形成一个开口，白本上的脚子线因为和祖本上的脚子线已经被——连接着，所以也同时形成了一个同样的开口，这时引入一根耳子线；这样，祖本的花就被传输到白本上了，成为行本或拼本。

The pattern draft copying technique is to copy another pattern draft according to the existing one. The basic principle of the pattern draft copying technique is as follows. First, the foot threads of the pattern draft that have been finished（Zu Ben）are connected with the foot threads of the one to pick（Bai Ben）one by one, and these foot threads are tied as X-shaped. Next, an ear thread of Zu Ben is pulled and lifted, and a foot thread floating on the ear thread is also lifted to form a shedding; for the foot threads of Bai Ben have been connected with that of Zu Ben one by one, a same shedding is formed at the same time. At this time, an ear thread is drawn in. In this way, the patterns of Zu Ben are transmitted to Bai Ben and become the Xing Ben（the pattern draft that can be operated on the loom）or the Pin Ben（the incomplete pattern draft used for pattern draft connection）.

首次挑花完成的花本，俗称祖本。祖本的脚子线和耳子线一般都采用较细的丝线，体积较小，便于保存。织机上使用的花本，必须将"祖本"通过倒花或拼花转换为脚子线和耳子线较粗的行本，行本即为可操作运行的花本。行本到机上再与牵线连接。具体步骤为：根据花本的大小，如该品种为一幅两花，则将牵线从中间平分为二份，然后分别伏在两个千斤鬲上。先将两千斤鬲平行挂在织机横梁的两边，然后将花本的脚子线与纤线连接，进行兜花。采用一根脚子线兜上两个千斤鬲的各一根牵线。直至兜完所有的脚子线，再将脚子线的两端——对结，形成线制环形花本，使花本可以在织机上连续循环使用。

The pattern draft picked for the first time is commonly known as Zu Ben. The foot threads and ear threads of the Zu Ben are generally thinner silk threads, which is small in size and easy to preserve. As for the pattern draft used on the loom, the Zu Ben must be converted into the Xing Ben with thicker foot threads and ear threads by the pattern draft copying or connecting techniques. The Xing Ben will be connected with the drawstrings on the loom. The specific steps are as follows. Firstly, the drawstrings should be divided according to the size of the Hua Ben. For instance, if the pattern draft of a certain brocade variety has two patterns on one piece, the figure harness should be divided into two halves, which are respectively laid on two Qian Jing Ge (literally, a heavy tube with a weight of one thousand jin). First, two Qian Jing Ge are hung on both sides of the beam in parallel, and then the foot threads of pattern draft and drawstrings are connected. One foot thread should be connected with one thread on each of the two Qian Jin Ge. Until all the foot threads are connected, the two ends of the foot threads are tied one by one to form a loop-shaped pattern draft, so that the pattern draft can be continuously and circularly used on the loom.

3. 拼花/Pattern Draft Connecting Technique

拼花就是把挑花或倒花制成的不完整的花本合并成一个完整的花本，使之具备上机织造的要求。

The pattern draft connecting technique is to combine incomplete pattern draft through the cross-stitch or pattern draft copying techniques into a complete pattern draft, so that it can meet the requirements of the loom weaving.

拼花可一人操作，也可数人同时操作，先将两本要拼的花本拴挂起来，分为上、下两本。操作时上层花本不动，在下层花本上作业。通俗地说，就是将上、下层花本的耳子线一一对应起来，把上层花本的耳子线引到下层，然后撤掉下层的耳子线。拼花的基本原理则是"同梭同色才能相拼"。拼好后还要对花本进行整理梳洗，以备上机织造。

The pattern draft connecting technique can be operated by one person or several people at the same time. First, two incomplete pattern drafts to combine are tied and hung up, one as the upper one, while the other as the lower one. When operating, the operator does not move the upper pattern draft, but only works on the lower one. In brief, the operator needs to correspond the ear threads of the upper and lower pattern draft one by one, draw the ear threads of the upper pattern draft to the lower one, and then remove the ear threads of the lower one. The basic principle of the pattern draft connecting technique is "only threads on the same shuttle and in the same color can be connected together". After the pattern draft connecting work, the operator should clean and comb the pattern draft for the loom weaving.

四、织机装造/The Assembly of the Draw Loom

织机装造俗称造机，就是根据所织锦的品种、规格，把织锦所需的经丝按地部组织、花部组织的不同要求分别安装到传统木织机的"内脏"（包括纤线、综线、综片、下柱、筘等）中，使其符合织造的需要。

The assembly of the draw loom, usually called loom building, means to install the warp threads with the different parts of the draw loom (including the drawstrings, the heald, the shaft, the lingoes, the reeds) according to the different requirements of ground and pattern fabric weaves, to satisfy the needs of brocade weaving.

1. 织机构造/The Structure of the Draw Loom

织锦机俗称花楼机。机架为木结构，部分零部件用竹材制作，只有少量的铁制品。其结构与《天工开物》所绘录的"花机图"基本一致。有两种机型：一种在云锦行业中称作"小花楼"机，拽花的位置在机右侧，称作"侧拽"（也称"横拉"）；另一种称作"大花楼"机，拽花的位置与织工相对，称作"对拽"（也称"竖拉"）。小花楼机与大花楼机不仅拽花的位置不同，在纤线结构、装造方法、拽花操作及适应生产的品种等方面都很不同。小花楼机只能织造纹样单位较小的提花织物，如云锦库锦类小花库锦。大花楼机即妆花织机，纤线较多，适合织大型的织物。

The draw loom is commonly known as the Hualou loom. Its frame is made of wood, some components are made of bamboo, and only a few components are made of iron. The structure of the draw loom is basically consistent with the Illustration of Patterning Loom recorded in *The Exploitation of the Works of Nature*, a compendium on industry, agriculture and artisanry written by the late Ming period scholar Song Yingxing. There are two types of draw looms. One is the small draw loom, or the small Hualou loom, called by the Yun brocade industry; the position for the thread puller is on the right side, and the way of drawing is called "side pulling", also known as "horizontal pulling". The other one is the big draw loom, or the large Hualou loom, and the position for the thread puller is opposite to that of the weaver, which is called "opposite pulling", also known as "vertical pulling". The small draw loom and the large draw loom are quite different not only in the thread pulling position, but also in the structure of drawstrings, the assembly method, the operation of thread pulling, as well as the brocade varieties produced. The small draw loom can only weave the jacquard fabrics with small pattern units, such as the Palace Brocade with small patterns. The large draw loom is the draw loom for Zhuanghua fabrics production, with more drawstrings, suitable for weaving brocade fabrics with large pattern units.

大花楼机长5.6m、高4m、宽1.4m，为斜身式，有坑机和旱机两种机型。坑机的前半身下部有一长2.8m、宽1.2m、深约0.5m的长方形地坑，过去民间一般多采用坑机，一是因为民间机房大多低矮，二是可以利用地坑的湿气便于织造。目前由于生产环境的改善，车间均为高房大屋，又有空调，所以绝大多数均用旱机。

The large draw loom is 5.6 meters long, 4 meters high and 1.4 meters wide, with an inclined shape. The large draw looms can be classified into two types: the pit loom and the dry loom. Under the front part of the pit loom is a rectangular pit with a length of 2.8 meters, a width of 1.2 meters and a depth of about 0.5 meters. In the past, most of the folk loom households used the pit looms, because many folk weaving workshops were low houses, and the moisture of the pit could facilitate the brocade weaving. However, at present, due to the improvement of production environment, workshops are mostly high houses with air conditioners, so the dry machines without pit are commonly used.

大花楼机由木质材料组成，采用榫卯、木楔连接，作用分工明确，结构受力合理，结实耐用。整个机身长达5.6m，从织口到机后经轴间参与织造工作的经丝长度为5.2m，经平线与水平线夹角为10°。提花纤线位置在经线长度的前3/7处，这样虽然提花动程达15～20cm，但其经丝的相对伸长仅为3%左右，经丝在纤线中的滑移很小，可以把经丝的伸长和摩擦控制在较小程度而利于织造。

The large draw loom is made of wood material, connected by mortise and tenon joints and wooden wedges, with components each with a clear function, reasonable structural stress and durability. The whole loom is 5.6 meters long, the warp thread participating in weaving from the cloth-fell to the warp beam behind the loom is 5.2 meters long, and the included angle between the warp line and horizontal line is 10 degrees. The drawstrings are placed at the first 3/7 of the warp threads, so that although the motion of pattern drawing reaches 15–20 cm, the relative elongation rate of the warp threads is only about 3%, and the slippage rate of the warp threads in the drawstrings is very small, which can help control the elongation rate and friction of the warp threads to a small extent, conductive to weaving.

大花楼木机的结构可分为机身、花楼、开口机构、打纬机构、送经卷取机构五大部分。各部机件名目繁多，由云锦艺人世代相传，多为口头俗名，不能尽证于古籍文献。清代晚期江宁人陈作霖的《凤麓小志》曾收录了当时南京"织缎之机名目百余"，同现今所用名目无多大差异。

The large draw loom consists of five main parts: the loom frame, the patterning tower, the shedding mechanism, the beating-up mechanism and the let-off and take-up mechanism, each part with many components under numerous names. Yun brocade craftsmen have passed these names down by words of mouth from generation to generation, and most of them are popular names, which cannot be all proved by ancient books and documents. *The Small Gazetteer of Phoenix Mountain*, written by Chen Zuolin, a native scholar of Jiangning in the late Qing Dynasty, recorded more than 100 names of satin weaving looms in Nanjing at that time, which were not much different from the names used today.

锦的妆花属于大花纹织物，其织物的开口运动分为地部和纹部两部分。地部是有规律的运动，而纹部却按照图案花纹的需要来运动，它们分别由范子、障子（就是现代织机的综框，管理经丝的升降）和纤线来管理。范子管地部组织，障子和纤线管纹部组织。提升范子形成的开口织入地纬，拽提纤线使纹部经丝上升，同时踩落障子，将拽提的部分经丝按一定的规律回落至原来的位置，这时所形成的开口织入纹纬，即彩纬和金、银线等。范子、障子的开口运动由织工用踏杆（俗称脚竹）来控制，纤线则由拽花工拽提花本上的耳子线来控制。

Zhuanghua fabric belongs to the fabric with large patterns, and the harness motion of the fabric can be divided into two parts: the ground part and the pattern part. The former moves regularly, while the latter moves according to the needs of the patterns. The shedding movement of both the ground part and the pattern part are both managed by the lifting harness, the lowering harness (the lifting harness is also named Fan Zi or Qi Zong, while the lowering harness Zhang Zi or Fu Zong, which control the up and down of warp threads) and the drawstrings respectively. The lifting harness controls ground fabric weaves, while the lowering harness and the drawstrings pattern fabric weaves. The shed formed by raising the lifting harness is woven into the ground weft, and the drawstrings are pulled up to make the pattern warp rise; at the same time, the lowering harness is stepped down to return the pulled warp to its original position according to a certain rule, and the pattern weft is woven into the shed formed now, that is, colored weft, gold and silver threads, etc. The motion of lifting harness and lowering harness is controlled by the weaver with the treadles (commonly known as the foot bamboo), while the drawstrings by the thread puller through pulling the ear threads of pattern draft.

2. 造机/The Loom Building

根据云锦织造开口要求，造机时除地部组织按素机（不提花的织机）装造外，还需使经丝与花本连接作提花装造。整个装造工艺包括打范子、捞范子、范子吊装、柱脚制备、数丝、引纤、拾绞、捞筘等工序。

According to the shed requirements of Yun brocade weaving, the ground weave fabric is assembled the way as that of plain loom, and the warp threads and pattern draft are connected for the jacquard weaving. The whole assembly process includes the following processes: preparing the lifting harness and the lowering harness, hoisting the lifting harness and the lowering harness, preparing the lingoes, counting threads, drawing in threads, silk threads picking, reeding and so on.

打范子（障子）就是将丝线绕在硬木制作的框架上。由于范子和障子的型制、制作材料和方法相同，习惯上把范子和障子制备工艺统称为打范子、捞范子、范子吊装等，实际均包括障子制备在内。范子、障子上的丝线绕制数量和每台织机需用的片数，是根据织物品种和规格来计算确定的。一般的妆花品种有7片范子、8片障子。计算好每片范子、障子所管理的经丝数量后，每片范子、障子就按计算好了的数量绕制丝线。绕制时在范子、障子架的中间形成一个上下相互环套的两个套圈以便于穿过经线。

Preparing the lifting harness or the lowering harness is to wind the threads around a frame made of hardwood. The number of silk threads wound on the frame and the number of shafts needed for each loom are determined by the variety and specification of fabrics. For example, the Zhuanghua varieties generally need 7 lifting harness and 8 lowering harness. After calculating the number of warp threads managed by each lifting harness and lowering harness, threads are wound on the frame, and two overlapping loops are formed in the middle of the frame so as to draw in the warp threads.

范子、障子打好后，按上机的顺序排列好，把整好经的经轴放在后面，再按照织物组织规定的穿综顺序和方法，将经丝一一穿入范子、障子扣内，范子穿上扣，障子穿下扣。因为范子只升不降，而障子相反，只降不升。穿综完成后，连同经轴一道上机安装。

After the preparation of the lifting harness and the lowering harness, they are arranged according to the order on the loom first; next the warp beam is placed behind; then the warp threads are drawn into the loops of the lifting harness (the loop above) and the lowering harness (the loop below) one by one according to the healding order and method required by the fabric weave, because the lifting harness only rises while the lowering harness, on the contrary, only falls. After the healding work is completed, the lifting harness and the lowering harness will be installed on the loom together with the warp beam.

范子和障子上机吊装的一般要求是：范子梁应保持水平，两端距机身相当，根据经面的倾斜度从前到后逐步调整吊装高度，以与经面倾斜度相吻合，保证开口清晰。

The general requirements for hoisting the lifting harness and the lowering harness on the loom are as follows: the beam for the lifting harness should be kept horizontal, with both ends equal to the loom frame; the hoisting height should be gradually adjusted from front to back according to the inclination of warp face so as to match the inclination of warp face and ensure the clear shed.

范子和障子排列的次序是从机头到机尾的方向先排障子后排范子。脚竹则从机头到机尾，按一扇范子一根脚竹的次序排列。范子和障子吊装调试定位后，在两侧机身上安装厢板加以定位。

The lifting harness and the lowering harness is arranged from the loom head to the loom end, the lifting harness first and then the lowering harness. The treadles is also arranged from the loom head to the loom end, in the order that one lifting harness is matched with one treadle. After hoisting, debugging and positioning of the lifting harness and the lowering harness the boards are installed on both the loom sides for positioning.

地部组织造机完成后，接下来就是纹部组织的装造了。先将柱脚准备好，柱脚是用15mm×500mm的竹制圆棍做的，一端打有孔眼，每根重15g左右。每根柱脚用丝线穿入端部的孔眼结成50cm长的套圈，按每根纤线一根柱脚制备。柱脚的重量很有讲究，轻了坠不住纤线，易造成经线开口不清；重了拽花工太费力。如果拽花工每次拽起200根经线，那就需要3kg以上的力，一天下来很疲劳。

After the assembly for the ground fabric weave is completed, the next step is the assembly for the pattern fabric weave. First, the lingoes should be prepared, which is made of 15mm × 500mm bamboo round sticks with holes at one end, each weighing about 15 grams. Each lingoe is drawn into the holes at the end with silk threads to form a loop 50cm long. One drawstring should be matched with one lingoe. The weight of the lingoe is essential, neither too heavy nor too light. If it is too light, it can't hold the drawstrings and may lead to the unclear shed of the warp threads; if it is too heavy, the thread puller might feel exhausted. If the thread puller pulls up 200 warp threads at a time, the thread puller must meanwhile pull the lingoes with a total weight of over 3kg, in which case he or she must be very tired at the end of the day.

柱脚制备好后，按照纤线数和总经数，计算每根纤线应吊经线数，并按计算结果将经丝分组数出、编绞，此过程称为数丝。

After the lingoes are prepared, the number of warp threads to connect with each drawstring should be calculated according to the number of the drawstrings and the total warp threads, and then the warp threads are calculated, grouped and braided according to the calculation, which is called "counting threads".

引纤需三四人协同进行，甲坐在机肚经面下划柱脚、钩线，乙在机侧兜纤、递纤，丙在花楼上挂纤，丁协助分经（如三人操作，丁的工作可由乙兼），纤线与脚子线对接示意图如图4-11所示。

It needs three or four people to cooperate on connecting the fiber harnesses with the foot threads. A sits under the warp surface on the loom body to mark the lingoes and hook the threads; B winds and passes the drawstrings on one side of the loom; C hangs the drawstrings on the pattern tower; and D assists in the warp dividing (if there are only three people, D's work can be currently performed by B). The drawstrings and the foot threads are connected as shown in Figure 4-11.

拾绫也称兜花，其方法与倒花工艺的兜花相同（图4-12），即是将花本与纤线连接的作

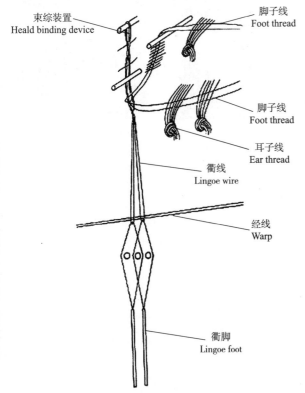

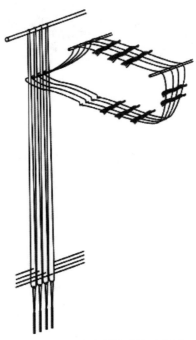

图4-11　纤线与脚子线对接示意图
The Schematic Diagram of Connecting the Drawstrings with Foot Threads

图4-12　花本与纤线连接示意图
The Schematic Diagram of Connecting Hua Ben with the Drawstrings

业，如图4-13所示。

Silk threads picking, also named pattern winding, is operated the same as that of the pattern winding technique（Figure 4-12）, that is, the operation of connecting pattern draft with the drawstrings, as shown in Figure 4-13.

捞筘也称穿扣，即将经丝按织物设计要求穿入扣齿，然后嵌于筘座上，如图4-14所示。

Reeding, also called denting, is drawing the warp threads into the reed-dent according to the fabric design requirements, and then embedding the reed-dent with silk threads on the sley, as shown in Figure 4-14.

以上是大花楼织机造机的一般过程和方法。在具体操作时，需视不同品种规格、纹样章法、结构等情况变化而灵活运用。

The above is the general assembly process of the large draw loom. In the specific operation, the looms should be assembled flexibly according to the changes of varieties, specifications, patterns and fabric weaves.

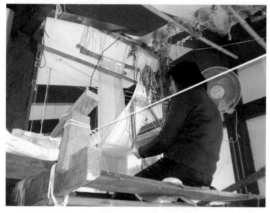

图4-13　兜花
Silk Threads Picking

图4-14　捞筘
Reeding

五、织造/The Weaving of Brocade

织锦木机完全人工织造，织工手足并用与拽花工相配合协同操作（图4-15）。足踏脚竹进行开口作业，手主要做投梭、纹刀引纬、过管、打纬等作业。

The wooden draw loom is operated manually, which calls for the cooperation of both a weaver and a thread puller（Figure 4-15）. The weaver should use both his or her hands and feet to operate the loom: the feet for shedding operation, while the hands mainly for shuttle picking, weft insertion with a wooden pattern stick, swivel weaving, beating-up and other operations.

引纬作业有三种方式，而织妆花品种时，这三种引纬方式都要用到。

There are three ways of weft insertion, and all these three ways should be used in Zhuanghua weaving.

一是投梭引纬，这与一般织物引纬相同，用梭子在经丝开口内往复运动将纬丝引入。用手工投梭，称为"甩梭"，如图4-16所示。

The first way is inserting weft with shuttles, which uses shuttles to reciprocate in warp shed to insert weft threads, the same as the weft insertion in the weaving of general fabrics. Shuttle picking with hands is also called "shuttle throwing", as shown in Figure 4-16.

图4-15 艺人操作大花楼织机
The Operation of the Large Draw Loom

二是纹刀引纬，用于引入片金。纹刀引纬，俗称铲纹刀。操作时右手将纹刀投入经丝开口，左手掐住一根片金后，右手将纹刀抽出，片金即留在梭口内，经打纬将其推入织口。

The second way is weft insertion with a wooden pattern stick, usually used to insert the gilt threads. When operating, the weaver puts the wooden pattern stick into the warp shed with the right hand, holds a gilt thread with the left hand, and then draws out the pattern stick with the right hand, so that the gilt thread will

图4-16 引纬织造
Weft Insertion

stay in the shed and be pushed into the cloth-fell by beating-up.

三是过管挖花。过管挖花操作时，需用纹刀作辅助工具。经拽花经丝形成开口后，将纹刀投入并翻转90°，使纹刀掌持梭口上下层经丝，这个动作称为"别纹刀"。然后手持纬管分段将该场次应织的色纬分别引入，这是用于"通经断纬"织法的引纬。所谓"通经断纬"即纬线由不定数的彩绒段拼接而成，不受颜色的限制。过管挖花又称挖花盘织，这种织法是云锦妆花织造工艺的一种独特方法（图4-17）。它可以根据需要在纬向同一梭内配织丰富多彩的彩纬，一般织七八种色，多的可达十几甚至几十种颜色。整匹的妆花织物中有几十朵主花，甚至可以配织成完全不同的色彩，真正做到了逐花异色。织制时各色纬只在花纹轮廓线内引入和中断，即所谓"断纬"。这不仅大大丰富了织品的色彩和艺术效果，而且色纬在地背无重叠，不增加织物的厚度，也节省了原材料。

The third way is inserting the warp through the swivel weaving operation and the pattern stick should be used as an auxiliary tool. After the thread pulling, the warp threads will form a shed. At that time the weaver puts the pattern stick into the shed and turns it over 90 degrees, so that the

图4-17 挖花盘织
The Swivel Weaving

pattern stick holds the upper and lower warp threads of the shed. This operation is called "turning the pattern stick". Next, the weaver holds the pirn to insert the color weft to be woven part by part. This way of weft insertion is used for the "passing warp thread and cutting weft thread" weaving method, which means that the weft threads are spliced by an indefinite number of colored velvet threads, with no limitations in color matching. The swivel weaving is a unique skill of Zhuanghua brocade weaving (Figure 4-17). Through the swivel weaving skill, the weavers can weave color weft, usually of seven or eight colors, and sometimes of more than ten or even dozens of colors, in the same shuttle broadwise. Dozens of main flowers in the whole Zhuanghua fabric can be woven into different colors, and even the colors of adjacent flowers can be different. When weaving, the multicolor weft threads are only inserted and cut in the pattern outline, which is called "cutting weft". This not only greatly enriches the color and artistic effect of the fabric, but also does not overlap the color weft on the ground, does not increase the thickness of the fabric, and saves raw materials.

挖花盘织技法的出现弥补了以往锦缎配色只能分段配色形成色条的缺陷。挖花盘织是我国古代丝织提花技术的进一步发展，出土和传世的明代丝织品中已大量采用这种织法。挖花过程中，除了彩绒外，还夹金织银，或织入孔雀羽等特殊材料，使织物显得雍容华贵，金翠交辉，富丽堂皇。

The swivel weaving technique improved the situation that brocade color matching can only be realized by sections to form the color stripes in the past. It is a further development of silk jacquard technology in ancient China, which has been widely used in those silk fabrics unearthed and handed down from the Ming Dynasty. In the process of swivel weaving, besides colored velvet, other special materials such as gold or silver threads, peacock feathers and so on, are woven into the fabric, which makes the fabric look elegant and magnificent.

在具体操作时，还要根据不同品种的织造工艺要求合理选择和安排基本操作，制订织造操作程序加以实施。织锦艺人在生产实践中不断总结操作经验，用简练概括的语言编出许多操作口诀，这些口诀既是操作要领，也是操作技术规范。这些历代传下的手工操作口诀总结了千百年来艺人们的经验，表现了织锦这一传统工艺充满着无比旺盛的生命力。机上坐着的人，称作"拽花工"，只要按过线顺序提拽即可，相当于在敲计算机键盘；机下坐着的人，称作"织手"，他面前的织造面相当于计算机显示屏，根据拽花工提起的经线开口织造，妆金敷彩，就能织出五彩缤纷的织锦来。

During the operation, the weaving procedures should be reasonably selected and operated according to the specific requirements of different brocade varieties. Brocade craftsmen constantly sum up their experiences of production practice into many mnemonic rhymes in brief and concise language, which covers not only main points of operation, but also technical regulations. These mnemonic rhymes for manual operation handed down from past dynasties sum up the experience of craftsmen accumulated for thousands of years, and show that brocade weaving, as a traditional craftsmanship, is full of incomparably vigorous vitality. The person sitting high the pattern tower, or the "thread puller", only needs to pull the threads according to the thread passing order of the pattern draft, which is roughly similar to typing on the computer keyboard; the person sitting under the pattern tower, or the "weaver", before whom the weaving surface is roughly as the computer display screen, weaves according to the warp shed formed by the thread pulling, and inserts gilt threads or colored velvet. In this way, the colorful and exquisite brocade is woven.

第二节　近代织锦技艺/Brocade Weaving Craftsmanship in Modern Times

近代织锦技术是以法国织机匠师贾卡（Jacquard）发明的机械式提花机为起点的。当西方人对"丝绸之路"运来的中国丝织品赞叹不已时，中国的织锦提花机技术和纹制加工技艺也沿着这条路传入欧洲。1725年，法国纺织机械师布乔（B.Bouchon）尝试机械提花，从中国的挑花结本中二进位制原理得到启发和借鉴，采用"穿孔纸带"的提花方法，但他的发明并不实用。真正使用二进位制原理发明纹板提花机的是贾卡，他于1790年基本形成了机械提花机的设计构想，即使用纹板上的孔洞代替花本上的经纬组织点，至1799年，制成了具有整套的纹板提动机构、配置更为合理的脚踏式提花机，为纪念贾卡的贡献，这种提花机被称为贾卡提花机。发展至今天，称为机械式提花机。

The brocade weaving craftsmanship in modern times is based on the invention of mechanical jacquard loom by Joseph-Marie Jacquard, a French textile mechanic and inventor. When Westerners were amazed at the Chinese silk fabrics shipped from the Silk Road, Chinese technology of brocade weaving loom and pattern processing were also introduced into Europe along this road. In 1725, B. Bouchon, a French textile mechanic, tried out the mechanical jacquard weaving, which was inspired by the principle of binary system in Chinese cross-stitch work for forming decorative patterns, and adopted the jacquard weaving method with "punch tapes", but his invention was not practical. Joseph-Marie Jacquard was the one who actually invented the pattern card jacquard machine with the binary system principle. In 1790, he basically formed the design idea of mechanical jacquard machine, that is, using holes in the pattern card to replace the warp and weft weave points on pattern draft or the pattern sheet. By 1799, he invented a treadle jacquard machine with a complete set of pattern card lifting mechanism and a more reasonable configuration. In memory of Jacquard's contribution, this type of jacquard machine was named after him. Today, it has developed into the mechanical jacquard machine.

一、机械式提花机概述 /An Overview of Mechanical Jacquard Machine

（一）大提花织物概述 /An Overview of Jacquard Fabrics

1. 织物的分类 /The Classification of Fabrics

按纹样循环及织物表面纹样分类，织物可分为素织物、花织物。

According to the pattern repeats and the surface patterns, fabrics can be divided into two types: plam fabrics and patterned fabrics.

（1）素织物。应用基原组织构成表面素洁的织物。

Plain fabrics. A plain fabric refers to a fabric with plain and clean surface formed by original fabric weaves.

（2）花织物。分为小提花织物和大提花织物。小提花织物是指用变化组织及联合组织所构成的织物，在多臂机上织制；大提花织物又称纹织物，是指一个花纹循环的经纬丝线数很多且不能在多臂机上织制的织物，这种织物必须在提花机上织制。

Patterned fabrics. Patterned fabrics fall into two categories: dobby fabrics and jacquard fabrics. Dobby fabrics refer to the fabrics composed of fancy weaves and compound weaves, which are woven on a dobby loom; jacquard fabrics, also known as jacquard weaving fabrics, refer to the fabrics with a large number of warp and weft threads in a pattern repeat, which cannot be woven on a dobby loom but must be woven on a jacquard machine.

2. 大提花织物与其他织物的区别 /The Difference Between Jacquard Fabrics and Other Fabrics

（1）大提花织物与小提花织物的区别。小提花织物一般在踏盘织机或者多臂织机上形成花纹图案。由于这两种织机所控制的综框数比较有限（一般在 16～32 页），所以小提花织物组织循环或者花纹循环不是很大，花纹变化不多，花纹整体比较简单；而大提花织物在提花机上生产，提花机上的龙头通过综丝可以控制上千根甚至上万根的综丝，从而可以控制一个花纹循环的几千根甚至几万根经纱，这样在提花机上织出的织物的花纹图案组织循环比较大，花纹图案整体复杂，变化比较多。

The difference between jacquard fabric and dobby fabric. Dobby fabrics are generally woven on tappet looms or dobby looms. Because the number of heald frames controlled by these two looms is relatively limited (generally 16–32), its weave repeat or pattern repeat is not very large, and its patterns are relatively simple and not varied; while jacquard fabrics are woven on jacquard machines, on which the jacquard head can control thousands or even tens of thousands of hedges, thus controlling thousands or even tens of thousands of warp threads in a pattern repeat, so that the patterns of the fabric woven on the jacquard machine have larger pattern repeats, and the overall patterns are more complex and varied.

（2）大提花织物与印花织物的区别。印花织物是用染料或涂料以印刷方法在织物表面形成图案的织物，通过印花方式，可使织物的表面图案有各种形式的变化；而大提花织物表面的图案是通过织物的经纱和纬纱以一定的排列形式（如色经色纬排列）和一定的结构相互交织形成。

The difference between jacquard fabrics and printed fabrics. Printed fabrics are fabrics with patterns printed with dyes or pigments on the fabric surface. Through printing, the surface patterns of fabrics can change in various forms; as for the patterns on jacquard fabrics, their patterns are formed by interweaving warp and weft threads in a certain arrangement（such as the arrangement of color warp and color weft）and a certain fabric weaves.

3. 大提花织物的分类/The Categories of Jacquard Fabrics

大提花织物按组织结构分类，可分为简单大提花织物、复杂大提花织物。

According to the fabric weave, jacquard fabrics can be divided into two types: plain jacquard fabrics and compound jacquard fabrics.

（1）简单大提花织物。由一个系统（组）纬纱和一个系统（组）经纱交织而成的大提花织物。

Plain jacquard fabrics. The plain jacquard fabrics refer to the jacquard fabrics woven by interweaving a system（group）of weft threads and a system（group）of warp threads.

（2）复杂大提花织物。由复杂组织作为基础组织而构成的大提花织物，如经二重、纬二重等重组织结构，双层、三层等多层组织结构，毛巾、起绒、纱罗等组织结构。

Compound jacquard fabrics. The compound jacquard fabrics refer to the jacquard fabrics with the compound weaves as basic weaves, such as backed weaves including warp–backed weave, weft–backed weave, multi–layer weaves including two–ply or three–ply weave, and jacquard fabrics like towel weave , tufted pile weave and leno weave .

4. 大提花织物设计内容/The Design of Jacquard Fabrics

要经过品种设计、纹样设计、意匠、轧纹板、装造、试织等工序。

The jacquard fabrics are designed and woven through the working procedures of variety design, pattern design, pattern grid drawing, pattern card embossing, loom assembly and test weaving.

（二）提花机分类/The Classification of Jacquard Machines

（1）按纹织信息输入方式分类，可分为机械式与电子式提花机。机械式提花机如图4-18所示。

According to the input mode of jacquard information, the jacquard machines can be divided into the mechanical jacquard machines and the electronic jacquard machines. The mechanical

图4-18 单动式上开口机械式提花机
The Single-lift and Upper-shed Mechanical
Jacquard Machine

jacquard machine is shown in Figure 4-18.

（2）按开口类型分类，可分为上开口、中开口、下开口、半开口、全开口。

According to the types of shedding, the jacquard machines can be divided into five main categories: the upper-shed jacquard machine, the center-shed jacquard machine, the under-shed jacquard machine, the semi-open-shed jacquard machine and the open-shed jacquard machine.

（3）按刀箱的运动情况分类，可分为单动式与复动式。

According to the motion of the griffe, the jacquard machines can be divided into the single-lift jacquard machine and the double-lift jacquard machine.

（三）机械式提花机工作原理/The Working Principle of Mechanical Jacquard Machine

如图4-19所示，机械式提花机各构件从上到下依次为花筒、纹板、横针、竖针、托针板、首线及钩子、通丝、目板、中柱线、综丝、综锤。纹板有孔，相应位置的纹针提升，该纹针控制的经丝提升，形成上层梭口，为经组织点；纹板无孔，相应位置的纹针不提升，该纹针控制的经丝不提升，形成下层梭口，为纬组织点。

The parts name of a jacquard machine from top to bottom are cylinder, pattern card, needle, hook, needle plate, neck cord and neck cord hook, harness cord, comber cord, lingoe wires, heddle and lingoe, as shown in Figure 4-19. When there is a hole on the pattern card, both the figured hook at the corresponding position and the warp threads controlled by the figured hook are lifted to form a bottom closed shed, which is a warp interlacing point; when there is no hole on the pattern card, neither the figured hook at the corresponding position nor the warp threads controlled by the figured hook are lifted, forming a bottom shed, which is a weft interlacing point.

二、构件编号和纹针数选用/The Numbering of Jacquard Machine Parts and the Setting of Figured Hook Number

提花机装造是指使经丝受提花机控制并按照纹样与组织设计的要求做开口运动。在提花机上，为使装造工作中的穿挂吊接不搞错，必须将横针、竖针、通丝等各构件编号，编号顺序根据意匠图来定。意匠图决定了每一块纹板相应位置的纹孔是轧还是不轧，是否轧孔决定了相应的纹针是否提升，纹针的提升与否决定了下挂的经丝是否提升，也就决定的织物的相应位置是经组织点还是纬组织点，从而构成了整个大提花织物。

The assembly of jacquard machine refers to the work of making warp threads controlled by the jacquard machine and shedding according to the requirements of pattern and fabric weave design. On the jacquard machine, in order to make sure that there is no mistake in the tie–up during the assembly, the parts such as needles, hooks and harness cord must be numbered, and the numbering sequence shall be determined according to the pattern grid. The pattern grid determines whether a hole at the corresponding positions on each pattern card is punched; whether the hole is punched determines whether the corresponding needles are lifted; whether the needles are lifted determines whether the warp threads hanging down are lifted and whether the corresponding position of the fabric is a warp interlacing point or a weft interlacing point. Thus, the whole jacquard fabric is formed.

左、右手机械式提花机的区别是，右手机械式提花机的开关手柄在右侧，花筒在左上方，花筒顺时针转动；左手机械式提花机的开关手柄在左侧，花筒在右上方，花筒逆时针转动。

The differences between left–handed and right–handed mechanical jacquard machine are as follows: the former with the switch handle on the right side, the cylinder on the upper left and rotating clockwise; the latter with the switch handle on the left side, the cylinder on the upper right and rotating counterclockwise.

（一）机械式提花机各构件编号/The Number of Each Component of the Mechanical Jacquard Machine

以右手机械式提花机为例，各构件编号如图4–19所示。

Taking the right–hand mechanical jacquard machine as an example, the number of each component is shown in Figure 4–19.

（1）意匠图次序。意匠图中，一个纵格代表一根纹针经丝），纵格的次序从右到左，即意匠图右侧第一纵格代表第一根纹针，管理每花左侧的第一根经丝。一个横格代表一块纹板（纬丝），横格的次序从下到上，即意工兵图最下面第一横格代表第一块纹板，管理织造中的第一根纬线。

The sequence of the pattern grid. In the pattern grid, a vertical grid represents a figured hook（a warp thread）, and the order of the vertical grid is from right to left, that is, the first vertical grid on the right side represents the figured first hook and manages the first warp thread on the left side of each pattern. A horizontal grid represents a pattern card（a weft thread）, and the order of the horizontal grid is from bottom to top. The first horizontal grid at the bottom represents the first pattern card and the first weft thread in weaving.

（2）纹板孔次序。从机前看，花筒在左侧，纹板首端在机前。纹板孔次序为自右而左，自上而下。

The sequence of punched holes in the pattern card. Seen from the front of the machine, the cylinder is on the left, and the head end of the pattern card is in the front. The order of punched

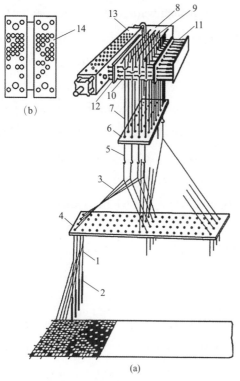

图4-19 单花筒提花机开口机构简图

A Schematic Diagram of the Shedding Mechanism
of the Single-cylinder Jacquard Machine

1—综丝/Heddle 2—综锤/Sinker 3—通丝/Harness cord
4—目板/Comber cord 5—首线/Neck cord
6—托针板/Base plate 7—竖针/Hook 8—刀架/Creel
9—提刀/Griffe 10—横针/Needle 11—弹簧/Spring
12—横针板/Needle plate 13—花筒/Cylinder
14—纹板/Pattern card

holes is from right to left and from top to bottom.

（3）纹针次序。横针与纹板孔相对应，机前最上一针为第一针，自上而下，自前向后；竖针与横针相对应，机前最左侧为第一针，自左向右，自前向后。

The sequence of needles and hooks. Needles correspond to the punched holes in the pattern card, and the uppermost needle in the front of the jacquard machine is the first needle, from top to bottom and from front to back; hooks correspond to the needles, from left to right and from front to back.

（4）目板孔次序。目板孔次序为自机前向机后，自左向右。

The sequence of holes in the comber board. The order of holes in the comber board is from forward to backward and from left to right.

（5）经丝次序。经丝次序为自左向右。

The warp sequence. The warp sequence is from left to right.

（6）纹板编连次序。纹板编连次序为1号→N号。

The lacing sequence of pattern cards. The pattern cards are laced and numbered from No.1 → No.N.

（二）提花机基本装造类型/The Assembly Ways of Jacquard Machines

目板上横向划分的区域称花区，纵向划分的区域称造。一个花纹循环中一根纹针控制的经丝数称把吊。提花机基本装造类型主要有以下几种。

The horizontally divided areas on the comber board are called the "pattern square", and the vertically divided areas are called "section". The number of warp threads controlled by a figured hook in a pattern repeat is called a "harness cord". The basic assembly types of jacquard machines mainly include the following ones.

（1）单造单把吊。单造是指纹针和目板在纵向只有一区。单把吊表示一根纹针在一个花区中只管理一根经丝。适用于经密小、花纹循环数小的织物。

The single-section and single-harness-cord assembly. Single section means that both the figured hooks and the comber board have only one pattern square lengthwise. Single harness cord means that a figured hook manages only one warp thread in a pattern square. The single-section and single-harness-cord assembly is suitable for fabrics with small warp density and small pattern repeats.

（2）单造多把吊。指一根纹针在一个花纹循环中管理两根或两根以上的经丝。又分为双把吊、三把吊和四把吊，双把吊最为常用。适用于经密较大或花纹循环数较大的织物。

The single-section and multiple-harness-cord assembly. Multiple-harness-cord assembly means that a figured hook manages two or more warp threads in a pattern repeat, including double-harness-cord, triple-harness-cord and quadruple-harness-cord. Double-harness-cord is most commonly used. The single-section and multiple-harness-cord assembly is suitable for fabrics with large warp density and large pattern repeats.

（3）双造。用于重经、双层、多层织物，几组经丝必须由不同的纹针管理。将纹针和目板的纵向分成相应区域称为分造。当甲乙两组经丝的比例为1∶1时，目板前后分成两个相等的区域，称双造。双造一般为单把吊，用于经密小、花纹循环数小的纹织物。

The double-section assembly. The double-section assembly recalls that several groups of warp threads must be managed by different figured hooks, used for the warp-backed, double-layer, multi-layer fabric fabrics. Dividing the areas lengthwise of figured hooks and comber board into the corresponding sections is called "dividing sections". When the ratio of warp threads between Group A and B is 1∶1, the front and back area of the comber board are divided into two equal sections, which is called "double-section". The double-section assembly is usually single-harness-cord, suitable for fabrics with small warp density and small pattern repeats.

（4）大小造。甲乙两组经丝的排列比不等时，纹针和目板纵向也要分成两个对应比例的区域，此时两造纹针数不等，称大小造。大造在前，单、双把吊均可；小造在后，一般为单把吊。

The major-and-minor-section assembly. When the arrangement ratio of warp threads in group of A and B is not equal, the figured hooks and the comber board should be divided into two corresponding areas in corresponding proportion lengthwise, and the number of figured hooks in two sections is not equal, which is called "major-and-minor-sections". The major section is in the front, with single-harness-cord or double-harness-cord; the minor section in the back, usually with single-harness-cord.

（三）纹针数的计算/The Calculation of the Number of Figured Hooks

纹针数的多少主要与花纹幅度（即"花幅"）、经密大小及把吊数有关。

The number of figured hooks is mainly related to the pattern width, warp density and the number of harness cords.

（1）单造单把吊。

The single-section and single-harness-cord assembly.

纹针数＝一个花纹循环经丝数＝内经丝数/花数＝花幅 × 经密

The number of figured hooks = the number of warp threads in a pattern repeat = the number of inner warp threads/the number of patterns = pattern width × warp density

（2）单造多把吊。

The single-section and multiple-harness-cord assembly.

纹针数＝一花经丝数/把吊数＝内经丝数/（花数 × 把吊数）＝花幅 ×（经密/把吊数）

The number of figured hooks =the number of warp threads per pattern/the number of harness cord = the number of inner warp threads/（the number of patterns × the number of harness cords）= pattern width ×（warp density/the number of harness cords）

（3）双造（或多造）。

The double-section（or multi-section）assembly.

纹针数＝一造纹针数 × 造数

The number of figured hooks = the number of figured hooks in one section × the number of sections

一造纹针数＝一花经丝数/造数＝内经丝数/（花数 × 造数）

The number of figured hooks in one section = the number of warp threads per pattern/the number of sections = the number of inner warp threads/（the number of patterns × the number of sections）

（4）大小造。大造可用单把吊，也可用双把吊。

The major-and-minor-section assembly. The major section uses the single harness cord or double harness cord.

大造纹针数＝大造内经丝数/（花数 × 把吊数）

The number of figured hooks for the major section = the number of inner warp threads in major section/（the number of patterns × the number of harness cords）

小造一般用单把吊。

The minor section usually uses the single harness cord.

小造纹针数＝大造纹针数 × 大造把吊数/大造比数

The number of figured hooks for the minor section = the number of figured hooks for the major section × the number of harness cords for the major section/the major section ratio

需要注意的是，要对纹针数进行修正；选用的纹针数应是地组织经丝循环数的倍数；纹针数最好是8或16的倍数。

It should be noted that the number of figured hooks should be revised; the number of figured hooks selected should be a multiple of the number of warp cycles of ground fabric weave; the number of figured hooks is preferably a multiple of 8 or 16.

（四）提花机规格 /The Specifications of Jacquard Machines

以号数（口数）来表示提花机规格大小，号数越大，纹针数越多。

The size specification of jacquard machine is expressed by Tex. The larger the Tex number, the more figured hooks.

花筒横向排列的孔眼称为列，大多为16列；花筒纵向排列的孔眼称为行。号数不同，行数有所不同。

The eyelets arranged horizontally in the cylinder are called "columns". Most of cylindes have 16 columns. The eyelets arranged longitudinally in the cylinder are called "rows". The number of rows varies with the Tex number.

为使纹板靠贴花筒时孔眼一一对准，在花筒每一小段上有两个铜栓，纹板对应位置有大孔，起到固定作用。因此，出现了整行（纹针数为16）与零针行（纹针数为14）。

In order to align the holes one by one when the pattern card is close to the cylinder, there are two copper bolts on each small section of the cylinder, and there are large holes at the corresponding positions of the pattern card, which play a fixing role. Therefore, there are full rows (with 16 figured hooks) and leftover rows (with 14 figured hooks).

实有纹针数=整行数×16+零针行数×14

The actual number of figured hooks = the number of full rows × 16 + the number of leftover rows × 14

如1400号提花机，实有纹针数为1480针。

For example, No.1,400 jacquard machine has 1,480 figured hooks.

三、辅助针的选用和纹板样卡设计/The Selection of Auxiliary Needles and the Design of Model Card

（一）辅助针的选用/The Selection of Auxiliary Needles

纹织物生产中，需要边组织和配备一些辅助装置，控制边纱和这些辅助装置的纹针，称为辅助针。辅助针主要有以下几种。

In the production of jacquard fabrics, it is necessary to form some selvedge weaves and equip with some auxiliary devices. The figured hooks that control the selvedge weaves and these auxiliary devices are called auxiliary needles. There are mainly the following types of auxiliary needles.

（1）边针。纹织物的边一般有内边和外边，外边也称小边或把门边。内边组织与纹织物地组织有关，把门边一般为平纹组织，由2~4根纹针控制；大边的纹针数与大边组织密切相关，针数为大边组织循环数的倍数。

Selvedge hooks. The selvedges of jacquard fabrics generally include the inside selvedge and outside selvedge. The inside selvedge weave is related to the ground weave of the jacquard fabrics, and the outside selvedge weave is generally plain weave, which is controlled by 2–4 figured hooks; the number of figured hooks in large selvedge is closely related to the large selvedge weave, and the number of figured hooks is a multiple of the number of the large selvedge weave cycles.

（2）棒刀针。棒刀片数与目板列数相等。棒刀针的多少取决于棒刀组织循环数。棒刀负荷较大，为保证起综安全，一般用2~3根纹针控制一片棒刀。棒刀针在辅助针中用量较多，常用的有32针、48针、64针、96针等。

Bannister shaft hooks. The number of bannister shafts is equal to the number of the comber board rows. The number of bannister shaft hooks depends on the bannister shaft weave cycles. The load of bannister shaft is heavy. In order to ensure the safety of shaft lifting, 2–3 figured hooks are generally used to control a bannister shaft. Bannister shaft needle is the most widely used type in all the auxiliary needles. There are 32 needles, 48 needles, 64 needles and 96 needles in common use.

（3）梭箱针。即控制梭箱升降的纹针，一般是单侧多梭箱用1~2枚针，双侧多梭箱用2~4枚针。

Shuttle box hooks. About the figured hooks controlling the shuttle box lifting, single–side multi–shuttle box is controlled by 1–2 needles, while double–side multi–shuttle box by 2–4 needles.

（4）投梭针。主要用于采用任意投梭机构的纹织物。为保证运转正常，一般每侧各用两枚针控制投梭机构。

Picking hooks. Picking hooks are mainly used for jacquard fabrics with any picking mechanism. In order to ensure normal operation, two needles are generally used on each side to control the shuttle mechanism.

（5）停撬针（停卷停送针）。用于抛梭织物上。特抛的纬纱间隔织入织物，在特抛花纹上形成纬二重结构，而在其他部分仍为单层结构。当特纬投入时，用纹针控制送经卷取机构，使其不产生送经和卷取运动（此纹针轧在特抛纬纱的纹板上）。

Regulator hooks (hooks used for stopping the take–up and let–off motion). Regulator hooks are used for woven fabrics. Extra wefts are inserted into the fabric at intervals, forming the weft–backed weave on extra patterns, while the other parts are still the single–layer weave. When an extra weft is inserted, the let–off and take–up mechanism is controlled by a hook to prevent the let–off and take–up motions (this hook is rolled on the pattern card of extra wefts).

（6）换道针。重纬纹织物的某些纬纱在织入一定距离后需要换色，俗称换道。用一枚针控制，在需要换色的前一纬纹板上轧孔。

Weft change hooks. Some weft threads of weft–backed jacquard fabrics need to change color after weaving at a certain distance, which is commonly known as "weft change". A hook should be used to control the weft change, which is rolled on the previous weft pattern card.

（7）色纬指示针。纬四重以上的纹织物，为了在处理停台时容易识别纬丝的颜色，每一色纬需用一枚针指示色别。

Color weft indicating hooks. For jacquard fabrics with more than triple weft, in order to easily identify the color of weft threads when processing stop, one hook should be used to indicate each color weft.

（8）起毛针、落毛针。毛巾织物在起毛圈之前及织平布之前各需要三枚针分别控制起毛圈和织平布。

Raising hooks and Dropping hooks. Three hooks are needed to control the terry motion and plain weaving respectively, before raising the terry towel fabrics and weaving the plain cloth.

（9）其他辅助针。控制不常用的特殊机构的针，如像景织物中控制前综管理接结经的纹针。

Other auxiliary needles. There are some other auxiliary needles that control special mechanisms not commonly used, such as hooks that control the front harness and manage stitching warp in photographic fabric weaving.

（二）纹板样卡的设计/The Design of Model Card

不同的织物，实际使用的纹针数及辅助纹针各不相同，因此必须对全部纹针进行合理安排，确定纹针、边针及其他辅助针的位置，称为纹板样卡，简称样卡。

Different fabrics actually use different figure hooks and auxiliary needles, so it is necessary to arrange all needles reasonably and determine the positions of figured hooks, selvedge hooks and other auxiliary needles, which is called the model pattern, or "model card" for short.

当织造某个纹织物的品种所需的纹针数和辅助针全部确定好以后，根据提花机各构件编号的要求和整机装造的需要，还要在提花龙头上对这些针进行合理的安排，确定纹针和辅助针的位置。

After the number of figured hooks and auxiliary needles required for weaving a certain kind of fabric are all determined, these needles and hooks should be arranged reasonably on the jacquard head to determine their positions, according to the requirements of the numbering of each jacquard machine part and the needs of the whole machine assembly.

样卡上的纹针数应等于实用纹针数。分造装造时样卡上须划出分造界线并指明造数。样卡上应标明各辅助针的位置。样卡正面首端一般用"号"表示，以免首尾、正反搞错。

The number of figured hooks on the model card should be equal to that of those practical. When dividing the sections, the boundary of dividing should be drawn and the number of sections should be indicated. The position of each auxiliary needle shall be marked on the model card. The head end of the model card face is generally indicated by numbers, so as to avoid confusion.

样卡设计应便于轧花操作和保护纹板，应注意以下几点。

The design of model card should be convenient for embossing operation and protecting the pattern card, and the following points should be noted.

（1）纹针有多余时，首先空出零针行，其次空出零针行周围的整行。

When there are spare figured hooks, the leftover rows should be emptied first and the full rows around the leftover rows.

（2）纹针尽量安排在中段，前后段应安排得较少，且前后段纹针数应尽量安排得相同。

The figured hooks should be arranged in the middle section as far as possible, with less arrangement in the front and back sections, and the number of figured hooks of the front and back sections should be arranged as much as possible.

（3）应用零针行时，一般最多用12针或8针，以保护大孔周围的牢度。

When leftover rows are used, 12 or 8 hooks are generally used at most to protect the fastness around the big holes.

（4）边针最好安排在纹板首端，不要夹在中间，以利于挂边时在机前可分左右。

The selvedge hooks are best arranged at the head end of the pattern card, not stuck in the middle, so that it can be divided left and right in front of the machine when hanging the selvedge.

（5）棒刀针的安排应使提花机负荷均匀，通常位于机前和机后，不能夹在中间，以免棒刀麻线夹起通丝。生产中常用2针或3针控制一片棒刀，以保证安全和减少单根纹针的负荷。

The bannister shaft hooks should usually be located in the front of and behind the machine, and cannot be stuck in the middle, so as to make the load of jacquard machine equally distributed and avoid the bannister shaft hooks clamping up the harness twine. In production, two or three hooks are often used to control a bannister shaft to ensure safety and reduce the load of a single figure hook.

（6）其他辅助针应安排在机后零针行。

Other auxiliary needles should be arranged in leftover rows behind the machine.

四、通丝计算与通丝穿目板/The Calculation of Hardness Cord and the Tie-up of Harness Cord on the Comber Board

提花机上直针带动经丝做单独升降运动的系统称为纹线。

The mechanic system on the jacquard machine that needles drive the warp threads to lift and drop independently is called "figure thread mechanism".

（一）通丝计算/The Calculation of Harness Cord

通丝上端套于首线钩，下端穿过目板，与综丝上的中（小）柱线连接。

The upper end of the harness cord is tied on the neck cord, and the lower end passes through the comber board and is connected with the medium (small) lingoe wires on the harness wire.

1. 选用通丝需注意的问题/Points for Attention in Selecting Harness Cord

（1）一台织机中，棉、麻通丝不能混用。

In a loom, the cotton harness cord and the linen harness cord can't be mixed.

（2）一台织机中，通丝的捻向要一致。

In a loom, the twisting direction of the harness cord should be consistent.

2. 通丝计算方法/The Calculation of Harness Cord

（1）通丝长度取决于提花机高度和所制织物的宽度。

The length of harness cord depends on the height of the jacquard machine and the width of the fabric.

（2）一台织机通丝总根数的计算。

The calculation of the total number of harness cord filaments in a loom.

单把吊装造的情况下，通丝把数＝纹针数；每把通丝根数＝花数。

In the single-harness-cord assembly, the number of harness cord bundles = the number of figured hooks; the number of harness cord filaments per bundle= the number of patterns.

多把吊装造的情况下，当把吊数为偶数时，每把通丝根数＝花数×（把吊数/2）；当把吊数为奇数，每把通丝根数＝花数×［1+（把吊数–1）/2］。

In the multiple-harness-cord assembly, when the number of harness cord is even, the number of harness cord filaments per bundle = the number of patterns ×（the number of harness cord/2）; when the number of harness cord is odd, the number of harness cord filaments per bundle = the number of patterns × [1 +(the number of harness cord–1)/2].

一台织机通丝总根数＝纹针数×每把通丝根数

The total number of harness cord filaments in a loom = the number of figured hooks× the number of harness cord filaments per bundle

（二）目板的作用及规格 /The Function and Specifications of the Comber Board

1. 目板的作用 /The Role of the Comber Board

目板的作用之一是供穿通丝用。装造时将每根通丝穿入一个目孔内，正确地隔离通丝，使其不紊乱。

One of the functions of the comber board is drawing in the harness cord. When assembling, each harness cord filament should be drawn into a comber board hole, and the harness cord filaments should be separated correctly so as not to be disordered.

穿目板还可以使经丝通按钢筘的幅度和密度均匀排列，以利于织造。

The tie-up of the comber board can also make the warp threads evenly arranged according to the width and density of the reed, which is beneficial to weaving.

2. 目板的规格 /The Specifications of the Comber Board

与经丝平行的目孔称行；与纬丝平行的目孔称列。

The comber board holes parallel to warp threads are called "rows"; and those parallel to weft threads are called "columns".

每10cm内有33行目孔，列数为55列。使用时需将多块目板根据品种镶拼起来。目板的拼接宽度为钢筘内幅+3cm×2或比织机内墙板小20～30cm。

There are 33 rows of holes and 55 columns in every 10 cm. When in use, a plurality of comber boards should be connected and assembled according to their varieties. The width of the connected comber boards is: the inner width of reed + 3cm × 2, or, 20–30cm smaller than the inner wallboard of the loom.

（三）目板的计算 /The Calculation of the Comber Board

1. 目板穿幅 /The Tie-up Width of the Comber Board

目板穿幅＝筘内幅或筘内幅+（1～2）cm

The tie-up width of the comber board = the reed inner width, or, the reed inner width +（1–2）cm

目板穿幅应安排在居中位置，然后根据花数画出每花宽度，各花交界处应空出两行目孔。若要采用棒刀，目板上要留出7行目孔的位置，供穿棒刀麻线用。

And the tie-up width of the comber board should be arranged in the middle, and then the width of each pattern should be drawn according to the number of patterns, and two rows of comber board holes should be vacated at the junction of each pattern. If a bannister shaft is adopted, 7 rows of comber board holes should be set aside for threading the bannister shaft twine.

2. **确定目板所穿行列数**/Determining the Number of Rows and Columns to Tie-up on the Comber Board

确定目板所穿行列数的前提是：目板穿幅与钢筘内幅相等，即每厘米目板下的经丝数等于钢筘中每厘米的经丝数。

The premise of determining the number of rows and columns of the comber board is that the number of warp threads per centimeter under the comber board is equal to the number of warp threads per centimeter in reed, because the tie-up width of the comber board is equal to the inner width of reed.

每厘米目板下的经丝数 = 目板每厘米行数 × 目板列数

The number of warp threads per centimeter under the comber board

= the number of comber board rows per centimeter × the number of comber board columns

应先确定列数，再计算行数。

The number of columns should be determined first and then that of rows.

（1）列数计算。

The calculation of the number of columns.

因为每厘米目板下的经丝数 = 每厘米钢筘内经丝数 = 目板行密 × 目板列数

Since: the number of warp threads per centimeter under the comber board

　　　= the number of warp threads per centimeter of reed

　　　= the row density of the comber board × the number of the comber board columns

所以初定列数（最小列数）= 每厘米钢筘内经丝数 / 目板行密 = 内经丝数 / （筘内幅 × 目板行密）= 筘号 × 筘入数 / 目板行密

The initial number of rows（minimum number of rows）

　　　= the number of warp threads per centimeter of reed/the row density of the comber board

　　　= the number of inner warp threads/(reed inner width × the row density of the comber board)

　　　= the reed number × the number of warp threads in each reed / the row density of the comber board

目板规格为每10cm 33行，计算时取目板行密为3.2行/cm。

The specification of the comber board is 33 lines each 10 cm, and the row density of the comber board is 3.2 rows/cm in calculation.

选取目板列数时，还应考虑以下因素：

①选定列数应为棒刀组织的倍数，当目板为分段飞穿时，各段列数应为飞穿数或筘入数的倍数；且列数应为把吊数的倍数，保证一综一孔的经丝密度。

②选定列数应大于初定列数。

③在可能的情况下，列数以少为好，以利于织造。常用20列、24列、32列、40列、48列，最多不超过50列，且尽量采用机前列数。列数还应避免采用不着16列，以免通丝摩擦厉害。

When selecting the number of the comber board columns, the following factors should also be considered: the selected column number should be a multiple of the bannister shaft weaves; when the comber board is skipped tied section by section, the column number of each section should be a multiple of the number of skipped ties or the number of warp threads in each reed; and the number of columns should be a multiple of the number of harness cord to ensure the warp density of one heald with one hole. The number of selected columns should be greater than the initial number of columns. If possible, the number of columns should be less, so as to facilitate weaving. 20, 24, 32, 40 and 48 columns are commonly used, with no more than 50 columns at most, and the number of columns in front of the machine should be used as much as possible. 16 columns should also not be used, so as to avoid severe friction of harness cord.

（2）行数计算。

The calculation of the number of rows.

每花目板实穿行数＝每花经丝数/选定列数＝内经丝数/（花数 × 列数）＝纹针数 × 把吊数/列数

The actual number of the comber board rows per pattern

= the number of warp threads per pattern/the number of selected columns

= the number of inner warp threads/（number of patterns × number of columns）

= the number of figured hooks × the number of harness cord bundles/ columns

每花实有行数＝花幅 × 行密＝花幅 ×3.2行/ cm

Actual rows per pattern = pattern width × row density = pattern width × 3. 2 rows/cm

因为选定列数＞初定列数，所以实有行数＞实穿行数，余行应均匀空出。

Since: the number of selected columns > the number of initial columns, the number of real rows is > the number of real rows, and the remaining rows should be evenly vacated.

（四）目板的穿向和基本穿法/The Basic Tie-up Methods of the Comber Board

各根通丝穿入目板上各个目孔的工作称为穿目板。穿目板是装造工作的重要环节。常用的穿法有一顺穿和飞穿。根据花纹形态和装造类型又有对称穿、混合穿及分造穿。

The work of drawing each harness cord thread into each comber board hole is called the tie-up of the comber board. Tying up the comber board is an important step in the loom assembly. The commonly used tie-up methods include straight-through tie-up and skipped tie-up. According to the pattern form and the assembly type, there are symmetrical tie-up, mixed tie-up and section-dividing tie-up.

我国习惯上采用的目孔排列顺序为：左边第一根经丝的目孔位置在目板第一列的左侧，经丝自左至右排列，目孔自前向后排列。

In China's textile industry, the comber board holes are most commonly arranged as follows: the hole for the first warp thread on the left is positioned on the left side of the first column of the comber board, the warp threads from left to right, and the comber board holes from front to back.

1. 目板的穿向/The Tie-up Direction of the Comber Board

目板的穿向是指通丝起穿目孔的位置和进行的方向。

The tie-up direction of the comber board refers to the starting point and direction that the harness cord is drawn into the comber board hole.

（1）顺穿向。通丝的起穿目孔为每花右后第一孔，从后到前，从右到左。最先穿目板的第一个通丝把挂于提花机右后侧最末一根纹针上，而最后穿的一个通丝把挂在机前第一根纹针上，即挂勾时，后穿先挂，如图4-20（a）所示。

Straight-through tie-up direction. The starting hole is the first hole to the right back of each pattern, tied from back to front and from right to left. The first harness cord thread to tie up first is hung on the last figured hook on the right back side of the jacquard machine, while the last thread is hung on the first figured hook in front of the jacquard machine, that is, tying up first and then hanging, as shown in Figure 4-20（a）.

（2）倒穿向。通丝的起穿目孔为每花左前第一孔，从前向后，从左向右。最先穿目板的第一个通丝把挂于提花机右后侧的纹针上，如图4-20（b）所示。

Reverse tie-up direction. The starting hole is the first hole to the left front of each pattern, from front to back and from left to right. The first thread bundle to tie first is hung on the figured hook on the right back side of the jacquard machine, as shown in Figure 4-20（b）.

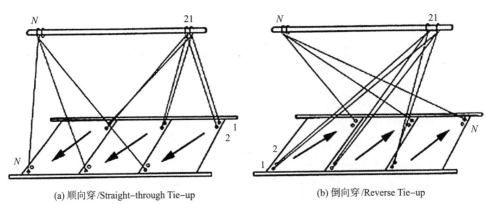

(a) 顺向穿/Straight-through Tie-up　　　　(b) 倒向穿/Reverse Tie-up

图4-20　目板的穿向
The Tie-up Direction of the Comber Board

2. 目板的基本穿法/The Basic Tie-up Method of the Comber Board

（1）一顺穿法：通丝按照目孔次序穿满一行后再穿第二行。优点是通丝摩擦小，操作简便。缺点是断经后找头不易。适用于经密较小的织物及双造或多造的每一造的目板上。

Straight-through tie-up. The harness cord is tied up in the first line according to the order of comber board holes, and in the second line. It has small friction of harness cord and simple operation as its advantages, and it not easy to find the thread head after the warp is broken as its disadvantage. It is suitable for fabrics with small warp density and for application on the comber board in each section of the double-section (or multi-section) assembly.

（2）飞穿法。将目板在一造内分成两段或多段，参照筘入数将通丝轮流穿入各段，穿满一行再换一行。对单把吊而言，飞穿数即为筘入数。

Skipped tie-up. The comber board is divided into two or more parts in one section, and the harness cord is tied up into each part in turn according to the number of warp threads in each reed. Another line cannot be tied-up until the previous one is finished. For the single-harness-cord assembly, the number of skipped ties is the number of warp threads in each reed.

例如，单把吊二段二飞穿，将目板纵向分成二段，通丝在段内连续穿二根后，跳至另一段再穿二根，依次前二后二跳穿。筘入数为飞穿数，等于2。

For example, as for the double-part and double-skipped-tie-up in a single-harness-cord assembly, the comber board is divided into two sections lengthwise; two harness cord filaments are tied up in a part first, and then jumps to another part in which two harness cord filaments are tied up, and, in this way, two filaments are tied up in turn in two parts. The number of warp threads in each reed is 2, equal to skipped ties.

对多把吊而言，为保持目板下"一孔一综"的密度，双把吊通丝应穿一空一，而三把吊通丝为穿一空一穿一。此时筘入数为经丝飞数而不是通丝飞数。优点是相邻两筘齿间经丝分离，断经过筘不会搞错。适用于经密较大的单造及大小造织物。

For the multiple-harness-cord assembly, in order to maintain the density of "one heald in one hole" under the comber board, in double-harness-cord assembly, the harness cord should be tied up according to the "one tie-up and one empty" principle, while in triple-harness-cord assembly it should be tied up according to the principle of "one tie-up, one empty and one tie-up". At this time, the reed number is equal to the warp counter instead of the harness cord counter. The advantages are that the warp threads between two adjacent reed dents are separated, and there is no mistake when the broken warp passes through the reed. It is suitable for the single-section or major-and-minor-section fabrics with large warp density.

（3）分造穿法。根据各组经丝比例，将目板纵向分成对应比例的区域，称为分造。目板都是先穿满一造后再穿另一造，先穿后造，再穿前造。在双造及多造中，各造比例相等，均采用一顺穿；若为大小造，则大造采用飞穿，小造采用一顺穿或二段一飞穿。穿筘时一组比例的经线穿入一个筘齿内。适用于重经、双层及多层织物。

Section-dividing tie-up. According to the proportion of warp threads in each group, the comber board is divided into areas with corresponding proportion lengthwise, which is called section-dividing. The comber boards are tied up section by section. In the double-section and multi-section assembly, since the proportion of each section is equal, the straight-through tie-up is adopted; in the major-and-minor-section assembly, the skipped tie-up is adopted in the major section, while the straight-through tie-up and one skipped-tie-up every two parts in the minor section. When reeding, a set of proportional warp threads should be tied into a reed dent. It is suitable for warp-backed, double-layer and multi-layer fabrics.

（4）对称穿法。当织物花纹为左右对称时，如被面、台毯等，为简化意匠，减少纹针，对称花纹的意匠只需画出一半，另一半由目板穿出。以对角线穿法为多。又可分为花边起穿和花芯起穿。花边起穿指通丝从目板两侧穿向中心。花芯起穿指通丝从目板中心穿向两侧。

Symmetrical tie-up. When the patterns of fabrics, such as such as quilt covers, table blankets, are bilaterally symmetrical, etc., in order to simplify the pattern design and reduce the figured hooks, only the half of the pattern grid with symmetrical pattern needs to draw, while the other half is tied up with the comber board.

以右手机为例，对称穿法的选择应根据意匠图画法来定。若意匠图为花纹的左半花，则目板采用花边起穿；若意匠图为花纹的右半花，则目板采用花芯起穿。其原因是若意匠图为左半花，则意匠图右面第一纵格为花芯，按右手机纹板号在机前，第一纹针所管理的经丝应按花芯要求运动。对右手机来讲，最末一把通丝挂于第一纹针上，且该经丝应排列于花纹中心，则目板穿法应采用花边起穿，先穿的通丝把挂于最末一根纹针上，最末一把通丝挂于第一纹针上。

Diagonal tie-up is most commonly used, which can also be divided into pattern-edge tie-up and pattern-center tie-up. The pattern-edge tie-up refers to tying the harness cord from both sides of the comber board to the center. The pattern-center tie-up refers to tying the harness cord from the center of the comber board to both sides.

若意匠图为右半花，第一纵格为花边，第一纹针下的经丝应按花边的要求运动，且该根经丝排列于花纹两边，挂于第一纹针下的为最末一把通丝，则最先穿的第一把通丝应从花芯开始，即花芯起穿。

Taking the right-hand loom as an example, the choice of symmetrical tie-up method should be determined according to the pattern grid. If the pattern grid is the left half of the pattern, the comber board is tied with the pattern-edge tie-up method; if the pattern grid is the right half of the pattern, it is tied with the pattern-center tie-up method. The reason is as follows. If the pattern grid is the left half of the pattern, the first vertical grid on the right is the pattern center. Since the pattern card number is in the front of the right-hand loom, the warp threads managed by the first figured hook should be moved according to the requirements of the pattern center; for the right-hand loom, the last bundle of harness cord should be hung on the first figured hook, and the warp threads should be arranged in the pattern center, thus the comber board should be tied with the pattern edge tie–up, the harness cord firstly tied up should be hung on the last figured hook, and the harness cord lastly should be tied up on the first figure hook.

需注意的是，花纹中心一根纹针所管理的两根经丝相邻且运动也一样，这种双经运动会使花纹中心变粗并破坏此处的地组织，称为并经现象。

If the pattern grid is the right half of the pattern, the first vertical grid is the pattern edge, and the warp thread under the first figured hook should be moved according to the requirements of pattern edge. The warp thread should be arranged on both sides of the pattern, and the last thread should be hung under the first figured hook, thus the first bundle of harness cord to tie up should be started from the pattern center, that is, the pattern–center tie–up.

为在花纹中心处仍保持单经运动，避免出现并经现象，在穿目板时应改进穿法。如图4-20所示，当为花边起穿时，剪去第 N 把通丝中左半区（倒穿区）的一根，顺穿区从1到 N 把，倒穿区从1到 N -1把，从第2目孔穿起，空目孔空在花边处；当为花芯起穿时，剪去第1把通丝中右半区（倒穿区）的一根，顺穿区从1到 N 把，倒穿区从2到 N 把，空目孔空在花边处。

It should be noted that the two warp threads managed by a figured hook in the pattern center are adjacent and move the same. This movement, called "double ends phenomenon", makes the pattern center thicker and destroys the ground weave here. In order to keep the single warp movement at the pattern center and avoid the double ends phenomenon, the tie–up method should be improved. As shown in Figure 4–20, when the pattern–edge tie–up is adopted, one of the left half area (reverse tying area) of the N th bundle of harness cord should be cut off, from 1 to N as the straight tying area, from 1 to N–1 as the reverse tying area, the second comber cord hole as the starting point, so that the vacant hole is left at the pattern edge; when the pattern–center tie–up is adopted, one of the right half area (reverse tying area) of the first bundle of harness cord, from 1 to N as the straight tying area, from 2 to N as the reverse tying area, so that the empty hole is left at the pattern edge.

对于边上剩下的目孔，可由相邻第二个通丝把上添一根通丝补入，称为补单（双）。是单是双看补通丝的那个通丝把所挂的纹针数是单数还是双数。

For the remaining holes on the edge, a harness cord filament can be added to the adjacent second harness cord bundle, which is called a single filament or double filament replenishing. Whether a single or a double filament should be replenished is determined by whether the number of figured hooks hung on the harness cord bundle to replenish are odd or even.

（5）混合穿法。在各种不同的花形处采用不同的穿法称混合穿法。中间自由花芯顺穿，两旁对称花采用对称穿，在两花形连接处，处理方法同"并经"。

Mixed tie–up. Different tying methods are adopted for different patterns, which are called "mixed tie–up". The pattern center in the middle is tied with the straight–through tie–up method;

the symmetrical patterns on both sides are tied with the symmetrical tie–up method; at the joint of two patterns, the treatment method is the same as that of "double ends phenomenon".

（6）零花穿法。为满足需要，达到一定的幅宽，可增加0.1～0.9花。当零花安排在机上右侧，则零花花形为左半花。当零花安排在机上左侧，则零花花形为右半花。由于零花位置不同，零花通丝穿目板时的方法也不同，需要特别注意。

"Leftover–pattern tie–up". In order to reach a certain width, 0.1–0.9 pattern can be added. When the leftover–pattern is arranged on the right side of the loom, the leftover–pattern shape should be the left half pattern. When the leftover–pattern is arranged on the right side of the loom, the leftover–pattern shape should be the right half pattern. Due to the different positions of leftover–pattern, the tie–up methods are also different, which requires special attention.

五、跨把吊装置及棒刀应用/Application of the Crossed–harness–cord Device and Bannister Shaft

单把吊适用于经密小、一花经丝数少的织物。当织物需加大花幅或增加经密但又不能暗加纹针数时，可采用双把吊或多把吊。当把吊数由单把吊变为多把吊时，织纹也会变粗。为了使地组织或间丝组织仍保持单经运动，可采用跨把吊或棒刀装置。

The single–harness–cord assembly is suitable for fabrics with small warp density and few warp threads per pattern. When the fabric needs to enlarge the pattern width or increase the warp density, but the number of figured hooks cannot be added, the double–harness–cord or multiple–harness–cord assembly can be used. When the number of harness cord bundles is changed from single to multiple, the fabric weave will become thicker. In order to make the ground weave or float binding weave still keep single– warp motion, the crossed–harness–cord device or the bannister shaft can be used.

（一）跨把吊 /The Crossed-harness-cord Assembly

1. 多把吊形式/The Multiple-harness-cord Assembly

采用双把吊时，通常采用下双把吊，即一根通丝下吊两根综丝，以减少通丝数。采用三把吊及以上时，需采用上下联合的形式，即每根纹针吊两根及以上的通丝，但每根通丝最多承受两根综丝的负荷。

As for the double–harness–cord assembly, the lower double–harness–cord is usually used, that is, two heald are hung under one harness cord filament to reduce the number of one harness cord filaments. When the triple (and more)–harness–cord assembly is adopted, it is necessary to combine the upper and the lower, i.e., each figured hook hangs two or more harness cord filaments, but each harness cord filament only bears the load of two heddles at most.

2. 多把吊的目板穿法及经丝穿法/The Tie-up Methods of the Comber Board and Warp Thread in the Multiple-harness-cord Assembly

多把吊时，目板列数应为把吊数的倍数，同一把吊的通丝不会分穿在两行目孔中。每根通丝下吊两根综丝，穿目板时通丝应穿一空一，保持"一综一孔"的经丝密度。

In the multiple-harness-cord assembly, the number of comber board columns should be a multiple of the number of harness cord bundles, and the filaments of the same harness cord will not be tied separately in two rows of comber board holes. Two heddles are hung under each harness cord filament; the harness cord filaments should be tied up according to the principle of "one tied and one empty", and warp density should be kept as "one heald and one hole".

3. 通丝穿目板的方法/The Tie-up Methods of the Comber Board

顺穿：一花中每根纹针下的通丝顺次穿入目板。

The straight-though tie-up: the harness cord filament under each figured hook in a pattern should be tied up into the comber board in succession.

跨穿：一花中每根纹针下的通丝交叉穿入目板。通丝的跨穿称为"上跨"。

The crossed tie-up: the harness cord filament under each figured hook in a pattern should be tied up with the comber board crosswise. The crossed tie-up of the harness cord filament is called "upper crossing".

4. 经丝穿综眼的形式/The Forms that Warp Thread Is Tied into the Heald Eye

顺穿：经丝顺次穿入通丝下的综眼中。

The straight-though tie-up: the warp threads are tied into the heald eyes under harness cord in sequence.

跨穿：经丝交叉穿入通丝下的综眼中。经丝的跨穿称为"下跨"。

The crossed tie-up: the warp threads are tied into the heald eyes under harness cord crosswise. The crossed tie-up of the warp threads is called "lower crossing".

在实际使用时，可将上跨、下跨联合使用，此时织纹最细致。图4-21所示为双把吊装造下经丝顺穿。

In actual use, the upper and lower crossing can be used together, and the fabric weave is the most delicate at this time. The straight-though tie-up of the warp threads in the double-harness-cord assembly is shown in Figure 4-21.

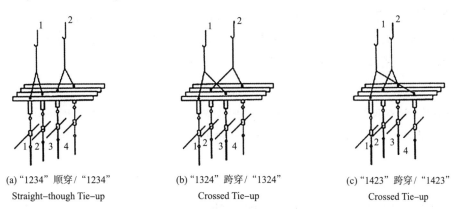

(a) "1234" 顺穿/ "1234"　　　　(b) "1324" 跨穿/ "1324"　　　　(c) "1423" 跨穿/ "1423"

Straight-though Tie-up　　　　　　Crossed Tie-up　　　　　　　　Crossed Tie-up

图4-21　普通双把吊和部分跨把吊结构

Common Double-harness-cord and Part of Crossed-harness-cord Structure

（二）棒刀 /The Bannister Shaft Device

1. 棒刀的结构和作用 /The Structure and Function of the Bannister Shaft

（1）多把吊织造时，用于分离同一把吊中的各根经丝，使其保持单经运动，相当于素机的综框。

In the multiple–harness–cord assembly, the bannister shaft can separate warp threads in the same harness cord bundles and make them keep the single warp motion, equivalent to the heald frame of the loom.

（2）有时单把吊织造时也用棒刀，当织物经密大时，综丝间相互牵连，易产生多少起疵点，这时棒刀不提升，仅起隔离作用。

Sometimes a bannister shaft is also used in the single–harness–cord fabric. When the warp density of the fabric is large, the heddles are easily intertwined with each other, which is easy to produce many defects. At this time, the bannister shaft does not lift, but only plays a separating role.

棒刀为狭长而薄的木片，厚4mm，高30~40mm，宽度大于筘幅宽。每列棒刀穿入同一列的中柱线圈环中，棒刀麻线穿过目板，吊挂于棒刀针下，由棒刀针带动棒刀片运动。

The bannister shaft is a long, narrow and thin wood chip, with a thickness of 4mm and a height of 30–40mm, and its width is larger than the reed space. Each row of bannister shaft passes through the lingoe loops of the same row, and the bannister shaft twine passes through the comber board and hangs under the bannister shaft hook, which drives the bannister shaft blade.

2. 棒刀组织的选择 /The Selection of the Bannister Shaft Weave

（1）棒刀的作用相当于综框，因此棒刀组织应选择有规律的纬面组织。

The function of the bannister shaft is equivalent to the heald frame, so the regular filling weave should be selected for bannister shaft weave.

（2）棒刀片数与目板列数应相等，且应为所起棒刀组织的倍数。

The number of bannister shaft is equal to that of the comber board columns, and should be a multiple of bannister shaft weaves.

3. 棒刀与纹针的配合 /The Match of Bannister Shaft and Figured Hooks

棒刀针控制棒刀，只能织造一些简单的纬面组织，如平纹、斜纹、缎纹组织，一般用作地组织。而花组织需棒刀与纹针共同配合。

Since the banniser shaft hook controls the bannister shaft, so it can only produce some simple filling weaves, such as plain weave, twill weave and satin weave, which are generally used as ground weaves. The pattern weave calls for the cooperation of bannister shaft and figured hooks.

有时纹针与棒刀提升规律相吻合，有时二者相冲而得到另一种新的组织。

Sometimes the lifting rule of the figured hooks are consistent with that of bannister shafts, while sometimes they may collide with each other to get another new weave.

六、意匠图绘制/The Drawing of Pattern Grid

纵格代表经丝（纹针），横格代表纬丝（纹板）。纵横格子的比例要与织物成品经纬密度之比相符合。因各种不同规格的织物经纬密度之比各不相同，故意匠纸也有很多规格。

The vertical grids represent warp threads (figured hooks) and the horizontal grids represent weft threads (pattern card). The ratio of vertical and horizontal grids should be consistent with that of warp density and weft density of finished fabrics. Because the ratio of warp and weft density for fabrics with different specifications is different, there are many specifications for pattern grids, too.

（一）意匠纸计算/Calculation of Design Papers

1. 意匠纸规格的计算/Calculation of Design Paper Specifications

我国常用的意匠纸规格有"八之八"到"八之三十二"共25种。前面数字代表横格数，后面数字代表与8个横格组成正方形时的纵格数。故"八之八"表示经纬密度相等，"八之十六"表示经密比纬密大一倍。大多数织物经密大于纬密；个别纬密大于经密的品种，可将意匠纸横用。

There are 25 specifications of design paper commonly used in China, from "8 × 8" to "8 × 32". The first number represents the number of horizontal grids, and the second number represents the number of vertical grids when forming a square with 8 horizontal grids. Therefore, "8 × 8" means that density of warp and weft are equal, and "8 × 16" means that the warp density is twice as large as the weft density. For most fabrics, warp density is greater than weft density. As for those fabrics with greater weft density than warp density, the design paper can be used horizontally.

意匠纸纵格次序从右到左，横格次序从下到上。纹织中常用组织都是8的倍数或约数，便于画意匠及轧纹板。

The vertical grids are arranged from right to left, and the horizontal grids from bottom to top. The commonly used weaves in jacquard weaving are multiples or approximations of 8, convenient for drawing design papers and punching pattern cards.

2. 意匠纸密度比的计算/Calculation of Density Ratio

选用意匠纸时，除考虑织物成品经纬密外，还要考虑织物的组织和装造情况。单经单纬纹织物意匠纸上每一纵（横）格代表一根经（纬）丝；重经织物一纵格代表甲乙一组经丝；双把吊或分造穿时一纵格代表把吊数或分造数的经丝。

When choosing the design paper, besides considering the warp and weft density of the finished fabric, we should also consider the weave and assembly of the fabric. On the design paper for single-warp and single-weft jacquard fabrics, each vertical (horizontal) grid represents a warp (weft) thread; on the design paper for the double-warp fabrics, it represents a group of warp threads of A and B; on the design paper for the double-harness-cord assembly or the section-dividing tie-up, it represents the number of harness cord bundles or warp threads in each divided section.

$$意匠纸密度比 = \frac{织物成品经密 / （把吊数 \times 分造数）}{织物成品纬密 / 纬重数} \times 8$$

$$\text{Density ratio of design paper} = \frac{\text{Warp density of finished fabric/ (The number of harness cord bundles} \times \text{The number of divided sections)}}{\text{Weft density of finished fabric/Weft number of finished fabric}} \times 8$$

算出的数值即为意匠纸规格"八之几"后面一个数值。意匠纸规格均为整数，计算所得若为小数时，可四舍五入取其整数，选用近似的意匠图。当计算所得结果小于8时，将分子分母颠倒（8不变），算出后，意匠纸横用。

The calculated value is the value behind "eight" in the specification number of design paper. The specifications number of design papers are all integers. If there are decimals in the calculation, they can be rounded to take their integers, and the design paper of approximate specification can be selected. When the calculated result is less than 8, the numerator and denominator should be turned upside down (8 remains unchanged), and the design paper of approximate specification is selected for horizontal use.

3. 意匠纸纵横格数的计算/Calculation of Vertical and Horizontal Grids on the Design Paper

意匠纸上纵格数与所用纹针数相同。当分造穿时，纵格数与一造纹针数相同；当分为大小造时，纵格数与大造纹针数相同。横格数是由纹样长度、纬密及纬重数决定的。纵横格数必须是地组织循环数的倍数，最好合大格整数，即8的倍数。

The number of vertical grids on the pattern grid is the same as the number of figured hooks used. When the section-dividing tie-up is used, the number of vertical grids is the same as that of figured hooks in one section; when the major-and-minor-section assembly is used, the number of vertical grids is the same as that of the figure hooks in the major section. The number of horizontal grids is determined by pattern length, weft density and weft number. The number of vertical and horizontal grids must be a multiple of the repeats of the ground weave, preferably a multiple of 8.

（1）纵格数计算。纵格数应为地组织循环数的倍数，使地组织能够连续；且纵格数还应合大格整数，以利于轧纹板。

Calculation of vertical grids. It should be a multiple of the repeats of the ground weave, and an integer equal to the number of large grids. The former is for the continuity of the ground weave, while the latter is beneficial to punching pattern cards.

单造单把吊：纵格数＝一花经丝数＝纹针数

For the single-section and single-harness-cord assembly:

The number of vertical grids = the number of warp threads per pattern = the number of figured hooks

单造多把吊：纵格数＝一花经丝数/把吊数＝纹针数

For the single–section and multiple–harness–cord assembly:

The number of vertical grids = the number of warp threads per pattern/the number of harness cord bundles = the number of figured hooks

双造或多造：纵格数＝一花经丝数/造数＝一造纹针数

For the double–section (or multi–section) assembly:

The number of vertical grids = the number of warp threads per pattern/the number of sections = the number of figured hooks per section

大小造：纵格数＝大造纹针数

For the major–and–minor–section assembly:

The number of vertical grids = the number of figured hooks in the major section

（2）横格数计算。

Calculation of horizontal grids.

横格数＝纹样长度 × 纬密/纬重数

The number of horizontal grids = the pattern length × weft density / weft number

修正：意匠纸横格数应为地组织和边组织纬丝循环数的倍数。

Correction: The number of horizontal grids of design paper should be a multiple of the repeats of the ground weave and selvedge weave.

（二）意匠图绘制步骤/The Steps of Drawing the Pattern Grid

1. 纹样放大/Enlarging the Pattern

纹样放大即将纹样移绘到已经计算好的意匠纸上。放样前，先在意匠纸上画好若干纵横格子范围，每一格子大小无统一规定，根据纹样粗细及放大技巧来定。纵横格配成正方形。格子不打线，用"×"表示其范围。分格次序从右到左，从下到上，最好在整幅内分成整数格。纹样划分成与意匠纸纵横格数相同的格子，移绘到意匠纸上对应位置。

Enlarging the pattern means that the pattern will be transferred and drawn on the calculated design paper. Before putting the pattern, a number of vertical and horizontal grids should be drawn on the design paper. There is no fixed regulation on the size of each grid, which is determined by the thickness of the pattern and the enlargement skills. The vertical and horizontal grids are matched into squares. The grids are only indicated by "×", without lineation. The grids are arranged from right to left, from bottom to top. It is best to divide whole design paper into grids of an integer. The pattern sample is divided into grids with the same number of vertical and horizontal grids, and transferred to the corresponding positions on the design paper.

2. 勾边/Pattern Bordering

绸面花纹的轮廓是由各根经丝的升降而形成的，因此必须把花纹的轮廓曲线转化为组织点曲线。勾边即依照纹样放大的铅笔线将占据小方格半格以上的涂格，不足半格的空出，均匀圆滑地涂出花纹组织点轮廓。

The outline of patterns on the satin surface is formed by the lifting and falling of each warp threads, so it is necessary to transform the curved lines of pattern into curved lines for weave interlacing points. Pattern bordering refers to painting the grids according to the pencil lines of enlarged patterns—to fill in those that occupy more than half a grid, and leave those less than half a grid empty, and evenly and smoothly paint the interlacing point.

需注意的是：意匠设色，不一定是纹样上所画颜色。花着色，地不着色，花上间丝点用另一种颜色。

It should be noted that the set color on the pattern grid is not necessarily the color painted on the pattern. The pattern should be colored, the ground should not be colored, and the float binding points on patterns should use another color.

（1）自由勾边。勾边时，只要沿着花纹轮廓勾得圆滑正确即可。适用于纹织物的花、地组织为斜纹、缎纹或其他不含平纹的变化组织，且不是跨把吊。

Free bordering. It is only necessary to delineate them smoothly and correctly along the outline of the pattern. Free bordering is suitable for those jacquard fabrics whose ground weave is twill, satin or other derivation weaves without plain weave, and not under a crossed–harness–cord assembly.

（2）平纹勾边。不论正织还是反织，意匠图上均为单起平纹。

Plain bordering. It requires all the odd number beginning plain weaves on the pattern grid.

单起平纹的经浮点在单数纵横格或双数纵横格相交处，双起平纹的经浮点在单数纵格和双数横格相交处，或反之。从一个经浮点过渡到另一个经浮点，其纵横过渡为奇数。平纹勾边可分为单起平纹勾边和双起平纹勾边。

For the odd number beginning plain weaves, the warp floating point is at the intersection of odd or even vertical and horizontal grids, and for the even number beginning plain weaves, the warp floating point at the intersection of odd vertical grids and even horizontal grids, or vice versa. The vertical and horizontal transition from one riser to another is odd number. Plain bordering can be divided into odd number beginning plain bordering and even number beginning plain bordering.

单起平纹勾边是指花纹勾边的起始点应点在单起平纹点上，即"逢单点单或逢双点双"，且纵横两个方向的过渡格数必须是奇数，以保证花纹的经浮长线在上下两端与平纹的纬浮点连接，不与平纹的经浮点连接，使花纹的轮廓不变形。适用于平纹地上起经花时的花纹勾边。

The odd number beginning plain bordering means that the starting point of pattern bordering should be on the odd number beginning point of plain weave, that is, "every odd number point or every even number point", and the number of transition grids in both vertical and horizontal direction must be odd, to ensure that the warp floating threads of patterns are connected with the sinkers rather than the risers of plain weave, and that the pattern contour lines will not be distorted. It is suitable for the pattern bordering to form the warp patterns on the plain weave ground.

同理，双起平纹勾边是指勾边起始点应双起，勾边时组织点的过渡格数为奇数。适用于平纹地上起纬花时的花纹勾边。可避免花纹的纬丝浮长线与平纹的纬浮点在左右方向相连。

Similarly, the even number beginning plain bordering means that the starting point of pattern bordering should be on the even number beginning point, and the number of transition grids for the interlacing points must be odd. It is suitable for the pattern bordering to form the weft patterns on the plain weave ground, which can avoid the connection between the weft floating thread and the sinkers of the plain weave in the left and right directions.

（3）变化勾边。由于跨把吊、大小造等装造形式，及某些组织结构的需要，意匠勾边过渡时，其纵横格数有一定要求，称为变化勾边。变化勾边一般有以下四种形式。

Variable pattern bordering. Due to the needs of the crossed-harness-cord or the major-and-minor-section assembly and some specific fabric weaves, there are certain requirements for the number of vertical and horizontal grids during the transition of pattern bordering, which is called "variable pattern bordering". There are generally four forms of variable pattern bordering.

第一种为横向偶数过渡，又称"双针勾边"。勾边时横向以1、2及3、4纵格为过渡单位，纵向可自由过渡。适用于两根纹针为单位的跨穿织物（1324及1423跨穿）、2：1的大小造织物等。当勾边时，横向要求以2、3及4、5偶数纵格为过渡单位时，又称"双针跨勾"，适用于某些起始位置变化的纬重平、方平组织。

The first is "double-hook bordering", i.e., horizontal grids of even numbers as the transition units. The vertical grids numbered 1 & 2, 3 & 4 are taken as transition units horizontally, and free transition can be made in vertical direction. It is suitable for crossed-tie-up fabrics with two figured hooks as units (1,324 and 1,423 crossed-tie-up), fabrics under the major-and-minor-section assembly with the ratio of 2：1, etc. In pattern bordering, the vertical grids numbered 2 & 3, 4 & 5 are taken as transition units horizontally, named "double-hook and crossed bordering", which is suitable for some double-weft plain weave and hopsack weave with different starting positions.

第二种为纵向偶数过渡，又称"双梭勾边"。勾边时纵向以1、2及3、4横格为过渡单位。适用于 $\frac{2}{2}$ 方平和经重平、表里纬之比为2：1的重纬花纹勾边。

The second is "double-shuttle bordering", i.e., vertical grids of even numbers as the transition units. The horizontal grids numbered 1&2, 3&4 are taken as transition units vertically. It is suitable for the pattern bordering of the $\frac{2}{2}$ hopsack weave and double-warp plain weave fabrics and the double-weft fabrics with the face and back weft ratio of 2：1.

第三种为纵横向均为偶数过渡，又称"双针双梭勾边"。适用于前两者综合因素下的花纹勾边。

The third is "double-hook double-shuttle bordering", i.e., both vertical and horizontal grids of even numbers as the transition units. It is suitable for pattern bordering under the comprehensive factors of the former two.

第四种为"多针多梭勾边"，当纵横过渡数为三格或三格以上的花纹勾边时采用。适用于大小造之比或表里纬之比为3∶1的织物勾边。

The fourth is "multi-hook multi-shuttle bordering", which is adopted when the vertical and horizontal grids as transition units are three or more. It is used for the pattern bordering of fabrics whose major and minor section ratio, or face and back weft ratio, is 3∶1.

3. 设色、平涂/Color Setting and Filling

在勾边前必须将各种花纹的颜色先设定好，称为设色。对意匠图中的图案轮廓进行勾边之后，必须将花纹轮廓所包围的部分用与勾边相同的颜色涂满，称为平涂。意匠图上的各种颜色只是代表不同组织结构，所用颜色要求色界分明。

Before pattern bordering, the colors of various patterns must be set first, which is called "color setting". After the pattern bordering for the contour lines in the pattern grid, the part surrounded by the contour lines must be filled with the same color as the bordering, known as "color filling". The various colors on pattern grids only represent different fabric weaves, and the colors used require clear color boundaries.

花、地采用不同组织，在意匠图上需用不同颜色涂绘。织物组织越复杂，意匠图上色彩也就越丰富。

For patterns and the ground, different weaves should be adopted, and different colors need to be painted on pattern grids. The more complex the fabric weave, the richer the colors on pattern grids.

4. 间丝点/Float Binding

在平涂的花纹块面上加上组织点，用来限制过长的经纱或纬纱浮长，这种组织点称间丝点。当经浮长过长时加纬间丝点，反之，纬浮长过长时加经间丝点。间丝点除限制经纬纱浮长外，还能增强织物牢度，提高花纹的明暗效果。间丝点一般分为平切间丝点、活切间丝点、花切间丝点三种。

To limit excessive warp or weft floating, interlacing points are added to the flat painted pattern surface, known as "float binding points". When the warp float length is too long, the weft float binding points are added; on the contrary, the warp float binding points are added when the weft float length is too long. In addition to limiting the warp and weft floating length, float binding points can also enhance the fastness of fabrics and improve the light and shade effect of patterns.Float binding points are generally divided into three types: plain interlacing points, free interlacing points and fancy interlacing points.

（1）平切间丝点。在意匠图上采用斜纹、缎纹或其他有规律的组织作为间丝组织，形成的间丝点称平切间丝点。它具有纵横兼顾的作用，即对经纬浮长都起限制作用。因此，在单层及重经、双层纹织物中应用较多。在重纬纹织物中，当花纹面积较大时也可应用平切间丝点。

Plain interlacing points. Twill, satin or other regular fabric weaves are used as the interlacing points for float binding on pattern grids, known as "plain interlacing points". It has both vertical and horizontal functions, that is, it limits both the warp and weft floating length. Therefore, it is widely used in single-layer, double-warp and double-layer fabrics. In double-weft jacquard fabrics, it can also be applied when the pattern area is large.

（2）活切间丝点。又称自由间丝点或顺势间丝点。在意匠图上依顺花叶脉络或动物的体形姿态点成间丝，这种间丝方法，既切断了长浮长纱线，又表现了花纹形态。活切间丝点一般只能切断单一方向的浮长，因此大多应用于重纬纹织物，而单层及重经纹织物也有少量应用。间丝点主要切断纬浮长。

Free interlacing points. Free interlacing points are formed according to the vein of flowers and leaves or the body posture of animals on pattern grids. This float binding method not only cuts off the long floating threads, but also displays the pattern shape. Generally, it can only cut off the floating length in one direction, so it is mostly used for double-weft jacquard fabrics, while sometimes it is also used for single-layer and double-warp jacquard fabrics. Free interlacing points mainly cut off the weft floating length.

（3）花切间丝点。又称花式间丝点，在意匠图上根据花纹的形状、块面大小等情况，将间丝设计成各种曲线或几何图形。这样除了能起到截断浮长的作用，还能够使花纹形态变化多样。花切间丝常以人字斜纹、菱形斜纹、曲线斜纹等斜纹变化组织为基础。

Fancy interlacing points. The interlacing points are designed into various curves or geometric figures according to the shape of patterns and the size of block on pattern grids, known as "fancy interlacing points". It can not only cut off the floating length, but also make the shape of patterns varied. Fancy interlacing points are often based on twill fancy weaves such as herringbone twill, diamond twill and curved twill.

5. 花、地组织处理/Treating Pattern and Ground Weaves

意匠图上绘花、地组织时，可分下列几种情况。

Pattern and ground weaves on the pattern grid can be treated in following ways:

（1）手工绘制意匠图时，当组织的组织循环数小于16，且为16的约数，组织为平纹、$\frac{1}{3}$ 斜纹、8枚及16枚缎纹等简单组织时，在意匠图上可以不必点出组织，只需在纹板轧法说明中说明组织的轧法。

For the manual drawing of the pattern grid, if the number of weave repeats is less than 16 and is an approximate number of 16, and the weave is simple, such as plain weave, $\frac{1}{3}$ twill, 8-heddle or 16-heddle satin weave, it is not necessary to mark the points on the pattern grid, but only to explain the punching method of the fabric weave in the instructions for punching pattern cards.

对于组织循环数大于16的复杂组织，或组织循环数小于16且不为16的约数的组织，以及泥地或变化组织等，必须在意匠图上全部点出组织。

However, for compound weaves with the number of weave repeats greater than 16, or less than 16 and not an approximate number of 16, crepe weaves and derivation weaves, all of points must be marked on the pattern grid.

（2）在采用纹织CAD编辑意匠图时，织物的花、地组织均可以铺在意匠图上，或者也可以不铺在意匠图上，用不同的颜色分别代表不同的组织即可。

For the drawing and editing of pattern grids with jacquard CAD, the pattern and ground weave can all be arranged on the pattern grid, or, different colors can be used to represent different weaves respectively, rather than be arranged on the pattern grid.

（3）当组织由棒刀织制时，意匠图上就不能点出组织，只需在纹板轧法说明中说明辅助纹棒刀针的轧法即可。

When fabric weaves are woven by bannister shafts, it is not necessary to mark the points on the pattern grid, but only to explain the punching method of the bannister shaft hooks in the instructions for punching pattern cards.

6. 建立纬纱排列 /Determining the Arrangement of Weft Threads

在意匠图绘制完成后，应确定纬纱排列比，以确定纬纱组数、一横格轧几张纹板、抛道信息等。

After the pattern grid drawing is completed, the weft arrangement ratio should be determined to fix the number of weft groups, the punching number of pattern cards for each horizontal grid, the extra weft information, etc.

7. 编制纹板轧法说明 /Writing the Instructions for Punching Pattern Cards

意匠图是纹板轧孔的依据。意匠图绘完后，必须编制纹板轧法以指导轧纹板工作。在纹织CAD中，轧法说明采用"组织表法"。手工绘制意匠图时，轧法说明一般采用"图示法""文字说明法"等。

The pattern grid is the basis for punching pattern cards. After drawing the pattern grid, it is necessary to write an instruction for punching pattern cards to guide the punching operation. In jacquard CAD, a "weave diagram" is used to explain the punching method. For the manual drawing of the pattern grid, the punching method is generally described by graphical illustrations or text explanations.

第三节	现代织锦技艺/Brocade Weaving Craftsmanship in Contemporary Times

现代织锦技术的标志是电子提花机的使用，电子提花机摒弃了传统机械提花机花筒和纸制纹板，将电子纹板的信号输入电子提花机的计算机中，直接控制纹针每一纬的提升。

The contemporary brocade technology is marked by the use of electronic jacquard machine, which abandons the traditional mechanical cylinder and paper pattern card, and inputs the signal of electronic pattern card into the computer for electronic jacquard machine to directly control the lifting of each weft on the figured hooks.

一、电子提花机工作原理/The Working Principle of the Electronic Jacquard Machine

（一）电子提花机的开口机构/The Shedding Mechanism of the Electronic Jacquard Machine

电子提花机的开口机构和机械提花机类似，也是由纹针带动经纱做单独运动，并形成梭口的纵向结构系统。

The shedding mechanism of the electronic jacquard machine is similar to that of the mechanical jacquard machine, which is also a longitudinal structural system in which warp threads are driven by figured hooks to move independently and form shed.

电子提花机与机械提花机的区别在于没有外在的纹板与花筒。如图4-22所示，电子提花机的纹板不是由像机械提花机的纸质纹板上对应的位置有孔还是无孔来表达的，而是通过纹板文件中的电子纹板相应位置有信号还是无信号来控制电子纹针的运动。因此电子提花机不存在横针与竖针，只有受电磁阀控制的纹针。

Its difference with the mechanical jacquard machine is that it has no external pattern card and cylinder. As shown in Figure 4-22, the pattern cards of electronic jacquard machine are not indicated by whether there are punched holes or not in the corresponding positions like mechanical jacquard machine, but the movement of electronic figured hooks is controlled by whether there are signals or not in the corresponding positions on the electronic pattern card in the pattern card file. Therefore, there is no horizontal needle and vertical hook in the electronic jacquard machine, and all the figured hooks are controlled by magnetic valve.

图4-22　电子提花机
An Electronic Jacquard Machine

（二）电子提花机织造过程/The Working Process of the Electronic Jacquard Machine

以博纳斯电子提花机为例，当电子纹板相应位置有信号，对应的电磁阀通电，对应的电子纹针提升，相应的经纱提升，织物上形成经组织点；反之，形成纬组织点。

Taking Bonas jacquard machine as an example, when there is a signal at the corresponding position on the electronic pattern card, the corresponding magnetic valve is powered on, the corresponding electronic figured hook is lifted, and the corresponding warp thread is lifted, too, thus the warp interlacing point is formed on the fabric. On the contrary, the weft interlacing point is formed.

二、电子提花机装造/The Assembly of the Electronic Jacquard Machine

（一）电子提花机各构件编号/The Numbering of Each Component of the Electronic Jacquard Machine

电子提花机各构件的编号形式与机械提花机各构件的编号形式有很多相同的地方。

The numbering of each component of the electronic jacquard machine has many similarities with that of the mechanical jacquard machine.

1. 意匠图顺序/The Sequence of the Pattern Grid

在电子提花机上生产纹织物时，意匠图都采用纹织CAD系统编辑，意匠图的纵格、横格次序要根据纹织CAD系统的设置而定。一般情况下，意匠图的纵格、横格次序设定为从上到下，从左到右，即意匠图最左边是第一个纵格，最上边是第一个横格。

When producing jacquard fabrics on the electronic jacquard machine, the pattern grids are edited by the jacquard CAD system, and the order of vertical and horizontal grids should be determined according to the setting of jacquard CAD system. In general, the order of vertical and horizontal grids is set as from top to bottom and from left to right; the first vertical grid is on the far left, and the first horizontal grid on the top.

2. 经纱顺序/Warp Sequence

一般情况下，电子提花机上的经纱顺序从左向右顺序排列，织机上的第一根经纱位于一个花纹循环的最左侧，即目板的一个花区的最左侧。

Generally, the warp threads on an electronic jacquard machine are generally arranged from left to right, and the first warp thread on the loom is located on the far left of a pattern repeat, that is, the far left of a pattern area on the comber board.

3. 电子纹针编号/The Numbering of Electronic Figured Hooks

电子提花机没有纹板，机上的电子纹针根据来自控制器的纹板文件的信号而做上下运动。电子纹针的排列顺序由电子提花机的控制器设定，通过修改程序可以改变电子纹针的编号。电子提花机的纹针排序通常有以下两种。

The electronic jacquard machine does not have a pattern card, and the electronic figured hooks move up and down according to the signal from the pattern card files of the controller. The arrangement sequence of electronic figured hooks is set by the controller, and the number of electronic figured hooks can be changed by modifying the program. Electronic figured hooks can be arranged in the following two ways:

（1）定左侧第一行的最后一个挂钩为第1针，从后向前在同一行中纵向编号，然后逐行顺排。最后一针为最右侧行的最前面一针。

The last hook in the first row on the left is set as the first one, and then the rest hooks in the same row are numbered longitudinally from back to front, and arranged row by row. The last hook is the front one of the far right row.

（2）定左侧第一行的最前一针为第1针，从前向后依次编号并逐行排列，最后一针为右侧末行的最后一个挂钩。

The front hook in the first row on the left is set as the first one, and then the rest hooks in the same row are numbered from front to back, and then arranged row by row. The last hook is the last one on the last row on the right.

4. 其他各构件编号/The Numbering of Other Components

连接第1根电子纹针的通丝为第1根通丝，穿第1根通丝的目板孔为第1目板孔，其余依次类推，如图4-23所示。

The harness cord which is connected with the first electronic figured hook is the first harness cord filament, and the comber board hole tied up with the first harness cord filament is the first comber board hole, and so on, as shown in Figure 4-23.

电子提花机为全开口梭口，一般情况下织物都采用正织，由因为没有外在的纹板，所以不需考虑纹板的编连方向和纹板作用纹针的次序（没有倒织与正织之分）。

The electronic jacquard machine is open-shed. Generally, the fabrics are all woven from bottom to top. Because there is no external pattern card, it is not necessary to consider the lacing order of the pattern cards and the sequence of figured hooks（there is neither the bottom-top weaving nor the top-bottom weaving）.

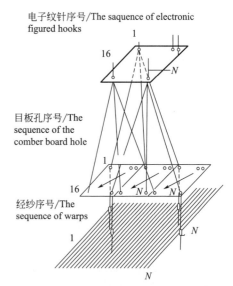

图4-23 电子提花机构件排序
The Sequence of the Electronic Jacquard Machine Components

（二）纹针样卡的设计/The Design of Model Cards

（1）纹针应合理安排在电子纹板的中间部位。

Figured hooks should be properly arranged in the middle of the electronic pattern card.

（2）每边余行数=（2688-纹针数）/（16×2）

The number of remaining lines per side =（2,688-the number of figured hooks）/（16×2）

（3）边针合理安排在首尾两端。

Selvedge hooks are reasonably arranged at both ends of the head and tail.

（三）通丝计算/The Calculation of Harness Cord

1. 通丝长度计算/The Calculation of Harness Cord Length

通丝长度取决于提花机高度和所织织物的宽度。

It depends on the height of the jacquard machine and the width of the fabric.

2. 一台织机通丝总根数计算/The Calculation of the Total Number of Harness Cord Filaments in a Loom

电子提花机的装造为单把吊装造，则：通丝把数＝纹针数，每把通丝根数＝花数；一台织机通丝总根数＝通丝把数 × 每把通丝数。

The assembly of the electronic jacquard machine is single–harness–cord, then:

The number of harness cord bundles = the number of figured hooks;

The number of harness cord filaments per bundle = the number of patterns;

The total number of harness cord filaments in a loom = the number of harness cord bundles × the number of harness cord filaments per bundle.

（四）电子提花机的目板设计与装造/The Design and Assembly of the Comber Board on the Electronic Jacquard Machine

1. 电子提花机的目板规格及计算/The Specification and Calculation of the Comber Board

电子提花机都属于高速提花机，故其目板大多用聚塑板制作，极其耐磨。目板列数一般取机上纹针列数，故目板常用列数有14列、16列、24列、32列。目板行密不设统一的规格，而根据织物的机上经密定制，没有多余的行数可供空余。电子提花机的目板纵深远小于传统织机的目不暇接板纵深，这有利于梭口的清晰和织机的高速运转。

The electronic jacquard machine belongs to the high–speed jacquard machine, so its comber board is mostly made of extruded polystyrene（XPS）board which is extremely hard–wearing. The number of comber board columns is generally taken from the number of figured hook columns, so the number of commonly used comber board columns is 14, 16, 24 and 32. There is no uniform specification for the column density of the comber board. The specifications are set according to the warp density of the fabrics, with no extra columns for vacancy. The longitudinal width of the comber board on the electronic jacquard machine is much smaller than that of the traditional machine, beneficial to the clear shed and the high–speed operation of the loom.

每花目板行密＝每花实穿行数 / 每花穿幅＝每花经纱数 /（列数 × 每花穿幅）

The comber board row density per pattern = the number of tie-up rows per pattern/the tie-up width per pattern = the number of warp threads per pattern/（the number of columns × the tie-up width per pattern）

制好目板后即可用色笔在上面画出穿幅、花区，如有零行应做好标记。

After making the comber board, the tie-up width and pattern area can be drawn with a color pen. If there are leftover rows, they should be marked as well.

2. 电子提花机的目板穿向/The Tie-up Direction of the Comber Board

电子提花机目板大多采用横向穿法，从后列起穿依次逐列（沿纬向）向前，分为左起穿和右起穿。左起穿的次序为从左到右，从后到前；右起穿的次序为从右到左，从后到前。

Most of comber boards on electronic jacquard machines are horizontally tied up. They are tied up from the back to front column along with the weft direction. According to the tie-up direction, it can be divided into the left tie-up and the right tie-up; the former from left to right and from back to front, the latter from right to left and from back to front.

3. 电子提花机的目板穿法/The Tie-up Ways of the Comber Board

电子提花机的纹针数较大，一般采用普通装造，即单造单把吊，没有多把吊和棒刀，所以目板穿法简单，大多采用横向一顺穿。但电子提花机有通孔板，要注通孔板和目板的穿法。

The number of figured hooks on the electronic jacquard machine is large, so the ordinary assembly, that is, the single-section single-harness cord, is usually adopted, without multiple harness cords and bannister shaft. Thus, the tie-up method of the comber board is simple, and the horizontal straight-through tie-up is most commonly used. However, the electronic jacquard machine has a through-hole board, so it is necessary to note the difference of the tie-up methods between through-hole boards and comber boards.

（1）通孔板的作用。通孔板所处的位置和传统提花机的托针板相同，但其功能是不一样的。通孔板的作用：一是使通丝相对于纹针只有上下作用力，使纹针挂钩在运动中不致晃动；二是开口时保持梭口清晰，这在阔幅织物中尤其必要。

The function of the through-hole board. The position of the through-hole board is the same as that of the traditional jacquard needle plate, but its function is different. The through-hole board has two functions: one is to make the harness cord only act up and down relative to the figured hooks, so that the hooks will not shake in motion; the other is to keep the shed clear during shedding, especially necessary in the weaving of broad fabrics.

（2）普通装造的通孔板、目板穿法。有通孔板的持花机，装造时应使穿通孔板孔和穿目板孔同时进行。新型提花机通丝挂钩采用弹性夹头，穿孔时把一排夹头从下向上压入通孔板孔后暂搁置在板上，然后把夹头下的通丝对应穿入各花的目板孔，从机后向机前逐排进行。

The tie-up method of the through-hole board and comber board in the ordinary assembly. For the jacquard machine with a through-hole board, the holes of the through-hole board and comber board should be tied up at the same time. The harness cord hooks of new jacquard machines adopt elastic chucks. During the tie-up operation, a row of chucks is pressed into the holes of the through-hole board from bottom to top and then temporarily puts on the board. Then the harness cord filament under the chuck is tied up with the comber board holes of each pattern, which is carried out row by row from back to front.

通孔板的孔和电子纹针上下对应，一般情况下，选目板列数和纹针列数相同，即目板列数和通孔板列数相同。目板的孔由于呈梅花状排列，每一列可看作两排，而通孔板由于孔直径较大，每一排又不得不分成两排，所以通孔板上的一列有四排，对应于目板的一列两排。

The holes of the through-hole board correspond to the electronic figured hooks up and down. In general, the number of comber board columns is the same as that of figured hooks, that is, the number of comber board columns is the same as that of through-hole board columns. Because the comber board holes are arranged in plum blossom shape, each row can be regarded as two rows; while the through-hole board has a large hole diameter, and each row has to be divided into two rows, so each column on the through-hole board has four rows, corresponding to one column and two rows of the comber board.

通孔板有两种穿法，一排顺穿法和两排联合穿法。目板也有两种穿法，顺穿法和跳穿法。

There are two tie-up methods for the through-hole board, that is, the one-row straight-through tie-up and the two-row combination tie-up. There are also two tie-up methods for the comber board, namely, the straight-through tie-up and the skipped tie-up.

通孔板和目板的穿法是互相关联的。当织高经密织物时，如果目板列数为32列，而纹针列数是16列时，通孔板采用一排顺穿法，目板也采用顺穿法；当目板列数和纹针列数相等且为16或32列时，通孔板采用一排顺穿法，目板采用跳穿法，或者通孔板采用两排联合穿法，目板采用顺穿法；当目板列数为16列，而纹针列数为32列时，通孔板应采用两排联合穿法，而目板采用跳穿法，通常在织宽幅但经密不高的棉织物时使用。

The tie-up methods of the through-hole board and the comber board are interrelated. When weaving high warp fabrics, if the column number of the comber board and the figured hooks are 32 and 16 respectively, the through-hole board adopts the one-row straight-through tie-up, and the comber board the straight-through tie-up; when the column number of both comber board and the figured hooks are all equal to 16 or 32, the through-hole board adopts the one-row straight-through tie-up, and the comber board the skipped tie-up, or the former the two-row combination tie-up, and the latter the straight-through tie-up; when the comber board column number is 16, but the figured hook column number is 32, the through-hole board adopts the two-row combination tie-up, and the comber board the skipped tie-up, usually used for the weaving of those broad cotton fabrics with low warp density.

三、意匠图绘制/The Drawing of the Pattern Grid

（一）纹织CAD编辑意匠图概述/Editing the Pattern Grid with Jacquard CAD

利用计算机进行意匠绘制和纹板轧孔的系统称为纹织CAD。采用纹织CAD进行意匠图绘制和生成纹板，效率得到极大的提高，故目前绝大部分提花织物生产厂家均采用纹织CAD系统进意匠图绘制。

The computer system used for drawing pattern grids and punching holes of pattern cards is called jacquard CAD. With the jacquard CAD system, the efficiency of drawing and generating pattern cards has been greatly improved, so most jacquard fabric manufacturers have adopted it at present.

纹织CAD系统主要包括三个方面：图像输入、图像与工艺处理、纹板输出。图像输入是指将纹样输入至纹织CAD系统中；图像与工艺处理是指图像的设计、编辑、色彩管理、纹织工艺处理等；纹板输出是指将纹板信息输出至电子提花机或纹板冲孔机中。

Jacquard CAD system mainly includes four parts: image input, image and technique processing, pattern card output. Image input refers to inputting patterns into jacquard CAD system. Image and technique processing refers to the design and edition of images, the color management, the jacquard technique processing, etc. Pattern card output refers to outputting pattern card information to the electronic jacquard machine or the pattern card punching machine.

（二）纹织CAD的一般应用流程/The General Application Process of Jacquard CAD

各种提花织物虽然是各有不同，但是在纹织CAD中的应用流程是基本相同的。

Although jacquard fabrics are different, their application process in jacquard CAD is basically the same.

（1）确定纹样大小。首先确定提花织物纹样的经纱和纬纱，然后确定经向循环宽度（cm）、纬向循环宽度（cm），即确定提花织物花回的大小。

Determining the pattern size. The warp and weft of jacquard fabric patterns are determined first, and then the warp repeat width（cm）and weft repeat width（cm），that is, the pattern repeat size.

（2）确立纹样经纬纱数。确定提花织物纹样的经纱密度（根/cm）和纬纱密度（根/cm）。

Setting the number of warp and weft threads of the pattern. The warp density（threads/cm）and weft density（threads/cm）of patterns on jacquard fabrics are determined.

（3）将纹样放入扫描仪。如果提花织物需要扫描输入，则将提花织物的纹样（布样、画稿等）按经线垂直水平的方向，正面朝下放人扫描仪中。扫描的大小就是提花织物花回的大小，如果花回太大，不能一次完成扫描，就需要将提花纹样分为若干个部分，依次扫描，最后将扫描的这若干幅图稿拼接在一起。

Putting the pattern into the scanner. If the jacquard fabric needs scanning input, its pattern（the cloth sample, the pattern draft, etc.）should be put into the scanner in the vertical face down in the warp direction. The scanning size is the size of the pattern unit. If the pattern unit is too large to complete scanning at one time, it is necessary to divide the pattern unit into several parts, scan them in turn, and finally put the scanned drafts together.

（4）选色、分色。图像扫描后，要对图像进行选色，选色之后对扫描图样进行分色，只需点击"分色"功能即可，软件会自动根据所选色进行分色。

Selecting and separating colors. After the image is scanned, it is necessary to select colors for the image. After the color selection, the color separation function needs to be clicked, and the software will automatically separate the colors.

（5）拼接。如果图像是分多次扫描，点击"拼接"功能先将这些图像拼接在一起。

Stitching image. If the images are scanned for several times, the stitching function can help put the images together.

（6）设置小样参数。打开"小样参数设置"对话框，在其中填入经纱密度、纬纱密度、经纱数、纬纱数这四个参数，其余的参数不用修改。

Setting the sample parameters. In the dialog box of the sample parameter setting, four parameters are filled in, including warp density, weft density, warp number and weft number, and the rest parameters need not be modified.

经纱数＝经纱密度 × 纹样花回宽度

The number of warp threads = the warp density × the width of a pattern unit

纬纱数＝纬纱密度 × 纹样花回高度

The number of weft threads = the weft density × the height of a pattern unit

（7）保存文件。设定好提花织物的小样参数后，就可以将初稿进行存盘，点击"保存文件"，将文件保存在指定的文件夹中。

Saving the file. After setting the sample parameters of the jacquard fabric, the first draft can be saved. "Save File" should be clicked, and the draft file will be saved in the assigned folder.

（8）修改图稿。保存好文件之后就可以对图稿进行修改了，可以充分利用绘图项中的相关工具对图样进行修改。修改时，是以织物中组织的种类来区分颜色的，即织物中的一种组织用一种颜色来表示，织物有多少种组织在最后的图样文件中就有多少种颜色。

Revising the draft. After saving the document, the draft can be revised. During revision, relevant tools in the drawing options can be made full use of. Moreover, the colors are distinguished by the types of fabric weaves. In short, one weave in the fabric is represented by one color, and there should be as many colors in the final file as the number of weave types in the fabric.

（9）组织分析。画好图稿之后，认真分析织物的每一种组织。可以将分析出的组织做好之后保存在CAD的组织库中。

Analyzing fabric weaves. After drawing the draft, every weave of the fabric should be meticulously analyzed. The analyzed weaves can be saved in the weave database of CAD.

（10）铺组织。将做好的组织铺人小样中，也可以不铺，在组织表中直接填入组织文件名。

Assigning fabric weaves. The completed weaves can be assigned in the sample; or, their names can be filled in the weave table directly, without weave assigning.

（11）生成投梭与保存投梭。根据提花织物的纬纱情况来确定投梭，然后将投梭保存。

Producing and saving the picking information. The picking information should be determined according to the weft of jacquard fabric, and then be saved.

（12）填组织表。根据织物的组织和意匠的颜色填写组织表。

Filling in the weave table. The weave table should be filled in according to the fabric weaves and the set colors in the pattern grid.

（13）建立样卡。根据织造当前提花织物的具体提花龙头的纹针吊挂形式建立样卡。

Creating a model card. The model card is established according to the form how the figured hooks needle is connected with the specific jacquard head to weave the current jacquard fabric.

（14）填辅助组织表。根据样卡中的辅助针，在辅助组织表中填出辅助针的组织。

Filling in the auxiliary weave table. According to the auxiliary hooks in the model card, their corresponding weaves should be filled in the auxiliary weave table.

（15）生成纹板。根据提花织物的类型、织机装造情况、提花龙头型号来选择具体的最后需要的纹板文件的类型，处理后即可以得到生产所需要的纹板文件。

Creating a pattern card file. According to the type of jacquard fabrics, the assembly of looms and the model of jacquard heads, the specific type of the final required pattern card file can be selected, and the pattern card file for production can be obtained after processing.

（16）纹板检查。对最后所得到的纹板文件进行检查。如果处理的是WB文件，可以利用纹板检查功能来分别检查单块的纹板；如果是其他类型的文件，则可以打开具体的文件类型来检查整体的纹板文件。

Checking the pattern card file. The final pattern card file needs to be carefully checked. If it is a WB file, each pattern card can be checked respectively with the pattern card checking function. If it is another type of file, the whole pattern card file can be checked by opening the specific file.

参考文献
References

［1］钱小萍. 中国织锦大全［M］. 北京：中国纺织出版社，2014.

［2］赵丰. 中国丝绸通史［M］. 苏州：苏州大学出版社，2005.

［3］四川丝绸公司. 四川丝绸史［M］. 成都：四川科技出版社，1990.

［4］唐林. 蜀锦与丝绸之路［J］. 中华文化论坛，2017（3）：20–25.

［5］陈显丹. 论蜀绣蜀锦的起源［J］. 四川文物，1992（3）：26–29.

［6］黄能馥，陈娟娟. 中国丝绸科技艺术七千年［M］. 北京：中国纺织出版社，2002.

［7］黄能馥. 中国成都蜀锦［M］. 北京：紫禁城出版社，2006.

［8］钟秉章，卢卫平，黄修忠. 蜀锦织造技艺［M］. 杭州：浙江人民出版社，2014.

［9］常璩. 华阳国志［M］. 重庆：重庆出版社，2008.

［10］吴方浪. 汉代"蜀锦"兴起的若干原因考察［J］. 丝绸，2015，52（9）：72–76.

［11］黄修忠. 蜀锦织造技艺［M］. 北京：化学工业出版社，2014.

［12］钱小萍. 丝绸织染［M］. 郑州：大象出版社，2005.

［13］黄修忠. 宋代时期的蜀锦技艺［J］. 四川丝绸，2007（4）：5.

［14］胡光俊，谭丹. 浅谈蜀锦及其传统织造技艺［J］. 现代丝绸科学与技术，2013，28（2）：51–54.

［15］赵丰. 中国丝绸艺术史［M］. 北京：文物出版社，2005.

［16］黄修忠. 浅谈蜀锦非物质文化的传承与发展［J］. 四川丝绸，2007（2）：52–54.

［17］王君平，王维. 蜀锦的代表产品及其生产工艺［J］. 四川纺织科技，2002（1）：58–60.

［18］沈干. 关于保护开发中国织锦的思考［J］. 纺织学报，2003（4）：99–102.

［19］杨长跃. 浅谈蜀锦传承与保护［J］. 四川丝绸，2007（4）：47–49.

［20］肖瑱，陈嘉，钟梦茹. "互联网＋"背景下南京云锦数字博物馆设计研究［J］. 西部皮革，2021，43（17）：137–144.

［21］穆弈君. 非遗云锦的动画实现与文创开发［J］. 艺术与设计（理论），2021，2（8）：110–112.

［22］王韦尧，张毅，杨丽. 宫廷审美视域下南京云锦的符号学探析与丝巾图案设计［J］. 丝绸，2021，58（6）：116–123.

［23］贾雪. 南京云锦博物馆导视系统设计改良策略研究［J］. 西部皮革，2021，43（4）：58–59.

［24］董泽昊，邓莉丽．云锦纹样在婚嫁文化创意产品设计中的应用［J］．美术大观（上），2021（2）：196-197．

［25］杨慧萍，李晴．基于云锦元素的系列文创产品设计研究［J］．才智，2021（1）：8-10．

［26］卢毅．传承教育语境下的非遗数字化传播探索——以南京云锦为例［J］．山东工艺美术学院学报，2020（6）：92-95．

［27］张庆善．对南京云锦保护的思考——从《红楼梦》中的江宁织造谈起［J］．中国非物质文化遗产，2020（2）：108-111．

［28］吴桐．传统云锦与现代织锦在织物纹样与组织结构方面的对比探讨［J］．轻纺工业与技术，2020，49（10）：69-70．

［29］吴婷．南京云锦作为文化符号的创新设计研究［J］．美与时代（上），2020（9）：30-32．

［30］钱小萍．中国宋锦［M］．苏州：苏州大学出版社，2011．

［31］随逍笑，姚鑫钰．"锦绣华年"系列服装设计作品［J］．丝绸，2021，58（9）：140．

［32］温润，王桂芳，吴思洋，等．基于生织匹染工艺的宋锦产品设计［J］．现代纺织技术，2021（6）：80-86．

［33］李岳．宋锦纹样在高级定制服装中的应用研究［D］．长春：长春工业大学，2020．

［34］张军，孙志芹．"宋锦"艺术特色及在中式服装设计中的应用［J］．艺术家，2020（12）：22-24．

［35］龙微微，徐萌．传统苏州宋锦纹样美感浅析［J］．美与时代（上），2020（9）：89-91．

［36］吴江统，芮美群．非遗宋锦的传承和创新［J］．华人时代，2020（8）：44-45．

［37］李春笑，王燕珍，曲洪建．宋锦文化认同对消费者购买意愿的影响研究——品牌认知的中介作用［J］．丝绸，2020，57（6）：11-17．

［38］李叶红．非遗"苏州宋锦"传承与创新设计专题的实践教学研究［J］．美与时代（上），2020（4）：130-131．

［39］梁丽萍．上久楷宋锦婚庆服装发布会展现中国传统文化内涵之美［J］．中国纺织，2019（11）：145．

［40］李易安．宋锦传承演绎下的地域文化研究［J］．江苏丝绸，2019（2）：18-21．

［41］罗炳金，赵秀芳．提花工艺与设计［M］．上海：东华大学出版社，2014．

［42］包振华．提花工艺与纹织CAD［M］．北京：中国纺织出版社，2015．

［43］翁越飞．提花织物的设计与工艺［M］．北京：中国纺织出版社，2004．